SEMICONDUCTORS AND SEMIMETALS

VOLUME 11

Solar Cells

Semiconductors and Semimetals

A Treatise

Edited by R. K. WILLARDSON

ELECTRONIC MATERIALS DIVISION
COMINCO AMERICAN INCORPORATED
SPOKANE, WASHINGTON

ALBERT C. BEER

BATTELLE MEMORIAL INSTITUTE
COLUMBUS LABORATORIES
COLUMBUS, OHIO

SEMICONDUCTORS AND SEMIMETALS

VOLUME 11   Solar Cells

Harold J. Hovel

THOMAS J. WATSON RESEARCH CENTER
IBM CORPORATION
YORKTOWN HEIGHTS, NEW YORK

1975

ACADEMIC PRESS   *New York   San Francisco   London*
A Subsidiary of Harcourt Brace Jovanovich, Publishers

ACADEMIC PRESS, INC.
111 Fifth Avenue, New York, New York 10003

United Kingdom Edition published by
ACADEMIC PRESS, INC. (LONDON) LTD.
24/28 Oval Road, London NW1

Library of Congress Cataloging in Publication Data

Willardson, Robert K   ed.
     Semiconductors and semimetals.

     Bibliographical footnotes.
     CONTENTS: v.   Physics of III-V compounds.—v. 3.
Optical properties of III-V compounds.—v. 5.   Infrared
detectors. [etc.]
     I.   Semiconductors.   2.   Semimetals.   I.   Beer,
Albert C., joint ed.   II.   Title.
QC612.S4W5   537.6'22   65-26048
ISBN 0–12–752111–9 (v. 11)

PRINTED IN THE UNITED STATES OF AMERICA
79 80 81 82   9 8 7 6 5 4 3

To my parents

Raymond and Kathe Hovel

Contents

Foreword

This book represents a departure from the usual format of *Semiconductors and Semimetals* in that the entire volume is devoted to a single article. Because of the importance of solar energy and the increasing amounts of research being done on solar cells, we feel that a detailed review is useful and timely. It is especially fortunate that Dr. Hovel, who has made numerous original contributions in this field, has been able to devote the time necessary to provide this valuable addition to our series.

The editors and author agreed to the use of cold type composition in order to reduce the production time to a minimum for this particular volume.

R. K. Willardson
Albert C. Beer

Preface

The importance of energy in our society has become all too apparent in recent years. Of the various energy alternatives, solar energy has the very desirable property of being essentially limitless and without the problems of pollution or physical danger. Solar cells are devices which convert sunlight directly into direct current electricity, and they have been an important part of the space program for over a decade. Solar cells are also capable of making a significant impact on terrestrial energy needs. It seems surprising then that there have been no books and very few review articles written about these devices, and it was this lack plus the author's interest in the subject that prompted the present volume to be written.

The book is intended as a background and general reference source, primarily for the nonexpert in the field of photovoltaics and those interested in entering the field. Solar cell experts may also find it useful, especially in its discussions of heterojunction and Schottky barrier cells, thin film devices, and polycrystalline devices. The book should be easily understandable for anyone who has had at least an introductory course in solid state electronics.

Chapter 1 is an introduction to solar cells and to the material and device parameters most important to these devices. The performance of a solar cell is equally dependent on its ability to generate photocurrent and its behavior in the dark. Chapter 2 describes the process of photocurrent generation and the spectral response, while Chapter 3 describes the electrical behavior in the dark. The efficiencies of Si, GaAs, and CdS solar cells under various conditions are discussed in Chapter 4. Since it would be impossible to describe the behavior of solar cells under all possible conditions, the trends in behavior caused by changes in junction depth, doping levels,

base and top region lifetimes, surface recombination velocities, presence or absence of aiding electric fields, etc. are
shown instead.

The effects of thickness on solar cell behavior are
discussed in Chapter 5, together with the effects of grain
boundaries in polycrystalline films. In Chapter 6, an introduction is given to Schottky barrier, heterojunction, vertical
multijunction, and grating solar cells. Radiation effects
on cells exposed to the space environment are discussed in
Chapter 7, and device behavior under various temperature and
intensity environments is described in Chapter 8. Chapter 9
serves as an introduction to solar cell technology, including
crystal growth, diffusion, ion implantation, antireflective
coatings, and Ohmic contacts.

I would like to express my appreciation for the many
helpful comments and reviews of the manuscript by my colleagues,
A. Onton, C. Lanza, R. Keyes, W. Dumke, W. Howard, B. Crowder,
P. Melz, and particularly J. Woodall. I am grateful to the
IBM Corporation for the encouragement and support given to me
during the undertaking. Most of all, I appreciate the patience
and understanding of my family during this seemingly endless
task.

Definitions

Air Mass	Secant of the sun's angle relative to the zenith, measured at sea level.
AM0	Solar spectrum in outer space.
AM1	Solar spectrum at earth's surface for optimum conditions at sea level, sun at zenith.
AM2	Solar spectrum at earth's surface for average weather conditions (technically, the spectrum received at earth's surface for 2 path traversals through the atmosphere).
Back Surface Field	A diffused or grown electrical contact to the base which blocks minority carriers but transmits majority carriers.
Base	The bulk region of the solar cell lying beneath the thin top region and the depletion region.
Dead Layer	A thin region adjacent to the front surface with very short lifetime. The dead layer is a result of the diffusion process, and extends over about a third of the top region.
Fill Factor	The fraction of the product of the short circuit current and open circuit voltage which is available as power output.
Inherent Efficiency	The power conversion efficiency of a solar cell without accounting for series resistance, shunt resistance, or reflection losses.
Open Circuit Voltage	Light-created voltage output for infinite load resistance.
Photocurrent	The current generated by light.

Short Circuit Current	The current for zero-net bias voltage across the device. It can differ from the photocurrent if a large series resistance is present.
Spectral Response, Absolute	The actual number of carriers collected per incident photon at each wavelength. Same as the quantum efficiency. Includes reflection from the surface.
Spectral Response, Relative	The number of carriers collected per incident photon at each wavelength, with the curve normalized to unity at the wavelength of peak response.
Spectral Response, External	The spectral response as it would be measured, including reflection of incident light. May be relative or absolute.
Spectral Response, Internal	The response of the solar cell without accounting for reflection from the front surface. May be relative or absolute.
Top Region	The thin, heavily doped region of the cell adjacent to the surface (i.e., the P region in a P/N cell or the N region in an N/P cell, etc.).

List of Symbols

A	diode perfection factor (in $\exp(qV/AkT)$)
A_a	active area
A_0	diode perfection factor in single exponential approximation
A_t	total area
A^*	Richardson constant for the dark current in Schottky barriers
A^{**}	A^* modified by scattering, tunneling, and quantum mechanical effects
B	argument in the exponential relationship for tunneling currents
C_0	surface concentration
D_n	diffusion coefficient of electron in p-type material
D_p	diffusion coefficient of holes in n-type material
d	thickness of antireflective coating
E	electric field
E_C	conduction band edge energy
E_F	Fermi level energy
E_g	bandgap energy
E_i	intrinsic Fermi level energy
E_{nn}	normalized electric field in p-type material
E_{pp}	normalized electric field in n-type material
E_V	valence band edge energy
E_r	energy level of recombination center
E_{av}	average energy of all photons in the source spectrum
$F(\lambda)$	incident photon density per second per unit bandwidth at wavelength λ
F_e	Fermi probability that a recombination center is occupied by a majority carrier
FF	fill factor
G	carrier generation rate due to incident light
H	total cell thickness (semiconductor regions only)
H'	cell thickness minus the junction depth and depletion width
I_m	current at maximum power point
I_0	dark current preexponential term

I_{00}	dark current preexponential term in single exponential approximation
I_{sc}	short circuit current
J	current density
$J_{dr}(\lambda)$	photocurrent density per unit bandwidth at wavelength λ due to collection from the depletion region
J_{inj}	injected current density
$J_n(\lambda)$	photocurrent density per unit bandwidth at wavelength λ due to electron collection from the p-side of the junction
J_0	dark current density preexponential term
$J_p(\lambda)$	photocurrent density per unit bandwidth at wavelength λ due to hole collection from the n-side of the junction
J_{ph}	photocurrent density (same as short circuit current density for negligible series resistance)
J_{rg}	space charge layer recombination current density in dark
J_{sc}	short circuit current density
J_{tun}	tunneling dark current
K_L	radiation damage coefficient
K_1	preexponential constant in tunneling current
K_1, K_2	fractions of the barrier heights on the two sides of a heterojunction
k	Boltzmann constant
L_n	electron diffusion length in p-type material
L_{nn}	effective diffusion length for electrons in p-type material when drift field is present
L_p	hole diffusion length in n-type material
L_{pp}	effective diffusion length for holes in n-type material when drift field is present
N_a	acceptor density
N_c	conduction band density of states
N_d	donor density
N_r	density of recombination centers
N_{ph}	total number of photons/cm^2 sec in the source spectrum
N_1, N_2	doping levels on the two sides of a heterojunction
N_v	valence band density of states
n	perfection factor in dark current of Schottky barrier
n	electron concentration
n	refractive index
n_i	intrinsic carrier density
n_n	electron density in n-type material
n_{n0}	electron density in equilibrium
n_p	electron density in p-type material
$n_{ph}(E_g)$	total number of photons/cm^2 sec with energies greater than the bandgap

P_{in}	input power density
P_R	number of recombination centers produced by each radiative particle
p_n	hole density in n-type material
p_{n0}	hole density in equilibrium
p_p	hole density in p-type material
p_{p0}	hole density in equilibrium
R	reflectance
R_s	series resistance
R_{sh}	shunt resistance
S_{back}	surface recombination velocity at back surface
S_{front}	surface recombination velocity at front surface
S_n	surface recombination velocity for electrons
S_p	surface recombination velocity for holes
T	transmission of light through metal film
T	temperature
V_d	built-in voltage
V_j	voltage across junction depletion region
V_m	voltage at maximum power point
V_{oc}	open circuit voltage
v_{th}	thermal velocity
W	depletion width
W_n	width of the n-region of a VMJ cell
W_p	width of the p-region of a VMJ cell
W_P	width of the lowly doped region in the base of a back surface field cell
W_{P+}	width of the highly doped region at the back of a back surface field cell
x_j	junction depth (width of the top region)
α	absorption coefficient
ΔE_c	conduction band energy discontinuity
ΔE_v	valence band energy discontinuity
$\Delta\phi$	image potential in Schottky barriers
ε_d	dynamic dielectric constant
ε_s	static dielectric constant
λ	wavelength
λ_0	wavelength for minimum reflection
ϕ	radiation fluence
ϕ_b	barrier height in Schottky barriers
ψ_b	low-high junction barrier height
ρ_{sh}	sheet resistivity
σ	capture cross section
τ	lifetime
μ_n	electron mobility in p-type material
μ_p	hole mobility in n-type material
θ	phase thickness of optical coating
χ	electron affinity

Semiconductors and Semimetals

CHAPTER 1

Introduction

A. Background

A solar cell is a photovoltaic device designed to convert
sunlight into electrical power and to deliver this power into
a suitable load in an efficient manner. The most important
application for solar cells in the past has been in the space
program. There are over 500 satellites of various types in
orbit around the earth, powered to a very large degree by
silicon solar cells, and it is safe to say that without these
cells, we would not have the sophisticated weather, communica-
tions, military, and scientific satellite capabilities we have
today. The advantages of solar cells lie in their ability to
provide nearly permanent, uninterrupted power at no operating
cost with only heat as a waste product, and their conversion
of light directly into electricity rather than some intermediate form of energy. They also have a high power/weight
ratio compared to other power sources. Their chief disadvantages lie in the low power/unit area of sunlight (that necessitates large area arrays), their relatively low efficiency,
and the degradation that takes place in hostile high energy
particle environments.

As useful as solar cells have been in the space program,
their potential importance for large-scale power generation
to meet earth's energy needs is even greater. A few years ago,
very few people would have seriously proposed solar energy as
a major power source; fossil-fuel burning, steam-powered generation plants were cheap and fuel supplies seemed inexhaust-
ible. Today, however, the rising price of fuel, the realiza-
tion that oil and gas supplies can only last a relatively few
decades, and the freedom of solar energy from pollution, have
all led to closer looks at solar energy as an alternative to
present day fossil-fuel systems.

1

One of the early schemes for large-scale power generation was the Arthur D. Little Company proposal to place large solar cell panels totalling about 100 km^2 in area into synchronous orbit about 36,000 km above the earth [1,2]. These panels would be capable of generating up to 15,000 MW of power, equal to 4 1/2% of the electric power used in the United States as of 1973. The panels would use readily available Si solar cells, which are about 11-12% efficient and highly radiation "tolerant." Other schemes for power generation via solar cells include large arrays of thin film CdS devices, which have received much attention in the past, and the use of either single crystal silicon "ribbons" or thin films of poly-crystalline silicon, which are just beginning to receive attention now.

The difficulty with all these schemes is their cost; if solar cells are to be competitive with other methods of power generation for terrestrial use, their cost must be reduced by a factor of several hundred. The prospects for significant cost reduction seem very good, particularly with the ribbon or polycrystalline Si schemes. The high cost of today's Si cells is in large part a result of stringent radiation toler-ance and starting efficiency requirements, as well as the absence of mass production methods for making the cells. The requirements for ribbon or polycrystalline Si devices operating on the earth's surface, on the other hand, would be greatly relaxed, and they should be readily adaptable to mass produc-tion with a minimum of problems. CdS solar cells are already made by low-cost techniques and are highly promising for use on earth, provided that certain degradation problems can be overcome (significant progress has been made on these problems in the last few years).

Virtually all solar cells now in use consist of a Si single crystal "wafer" 12-18 mil thick having a very thin (0.2-0.5 μm) diffused region at the surface to form a p-n junction (this surface region could also be produced by vapor growth). Electrical contact is made to the diffused region in such a way as to allow a maximum amount of light to fall on the Si; Ohmic contact is also made to the back of the wafer. A sputtered or evaporated antireflection coating is applied to reduce the amount of light lost by reflection from the surface. Finally, a "cover glass" of quartz, sapphire, or specially treated glass, with additional antireflection and ultraviolet rejection filters, is bonded to the cell with transparent adhesive with the purpose of preventing high energy radiative particles from reaching and degrading the device. The completed solar cell is shown in Fig. 1. The heavily doped side of the junction, produced by diffusion or vapor

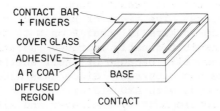

FIG. 1. Cross section of the common form of Si solar cell used in the space program in the past. Thicknesses are not to scale.

growth, is called the top region of the solar cell. The other side of the junction, consisting of the remainder of the substrate except for the depletion region, is called the base.

The question was raised in the late 1950's as to what the optimum material is for solar cell fabrication. Theory indicated that the optimum in terms of efficiency was a semiconductor with a bandgap of around 1.5 eV, although materials with bandgaps from 1.1 to 2.0 eV should be nearly as good. The chief candidates were Si, InP, GaAs, CdSe, and CdTe. Silicon quickly showed itself to be superior to any of the others experimentally. Of all the available semiconductors, Si is the most abundant, the least expensive, and the most technologically advanced. No material has yet proven to be superior to Si in converting outer space sunlight, although GaAs cells have come quite close. GaAs is far more expensive than Si and not nearly as technologically developed. However, GaAs solar cells are able to operate at higher temperatures and have greater radiation tolerance, making them attractive for some special applications.

The Si solar cell most widely used in the space program in the past has consisted of a 10 ohm-cm boron-doped wafer, diffused on one side to a depth of 0.2-0.5 μm with phosphorus, contacted with Ti-Pd-Ag or Ti-Ag-solder, and covered with 800 Å of SiO. These "N on P" devices can convert 11.5% of AM0 (outerspace) sunlight and 14% of AM1 (sea level at noon on the equator) sunlight into useful power. (Air mass is a term which describes how sunlight is modified by passage through the atmosphere.) This solar cell design is a compromise between AM0 efficiency and radiation tolerance; cells made from lower resistivity boron-doped substrates yield higher starting efficiencies but degrade under high energy particle radiation at a faster rate than the 10 ohm-cm devices.

Several substantial breakthroughs in Si cells have occurred in recent years. The first of these was the discovery
by Wysocki and co-workers [3] that P on N Si cells doped with
Li in the base exhibit greatly improved radiation tolerance
compared to other cells because of their ability to "recover"
after the damage has occurred. Since this discovery, Li-doped
cells have been made with AM0 efficiencies of 13.8%. The
second was the development by Lindmayer and Allison [4] at
Comsat Laboratories of a diffusion technology which eliminates
the heavily damaged (low lifetime, low mobility) so-called
"dead region" at the Si surface often caused by normal diffusion techniques. As a result of their work, cells with
enhanced response at blue and ultraviolet wavelengths have
been produced with AM0 efficiencies of 15.5%; the same cells
also exhibit improved radiation tolerance compared to the
conventional Si cells. The third significant development was
the discovery [5] that diffusing a heavily doped region at
the back contact to the cell raises the output voltage of thin
(<8 mil) Si solar cells; open circuit voltages of over 0.6 V
are obtained with such cells compared to the 0.55 V that would
normally be observed.

Significant advances have also been made in GaAs solar
cells in recent years. These devices are usually made by
diffusing Zn to a depth of around 0.6 μm into an n-type wafer
doped to $1-5\times10^{17}$ cm^{-3}, contacting with Ni or Ag, and covering
with SiO. In 1962 [6], such cells were around 9% efficient
at AM0 and 11-12% at AM1. In the last several years, the
application of liquid-phase epitaxial techniques for producing GaAs and $Ga_{1-x}Al_xAs$ layers has resulted in enhanced output
from GaAs-type cells [7-9], with AM0 efficiencies of 15% and
AM1 values of 19%. Progress for this type of solar cell is
still being made, but as yet the surface area of these laboratory devices remains in the $0.2-0.6$ cm^2 range (rather than
the 2-4 cm^2 areas of standard Si cells).

CdS solar cells have also shown progress recently. These
devices are made by evaporation of CdS onto metal foils or
metallized plastic, followed by immersion in a CuCl bath to
form a thin, highly conducting Cu_2S layer on the CdS surface.
They have suffered from degradation problems in the past, due
mostly to instabilities in the Cu_2S, but many of these problems have been minimized in the last several years, and stability projections of over 20 yr have been made for CdS cells
in controlled environments [10,11].

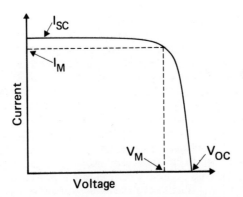

FIG. 2. *Voltage and current output from an illuminated solar cell.*

B.   Device Parameters

Solar cell behavior can conveniently be examined through three main parameters (as shown in Fig. 2): the open circuit voltage V_{oc}, which is the voltage output when the load impedance is much greater than the device impedance; the short circuit current I_{sc}, which is the current output when the load impedance is much smaller than the device impedance; and the "fill factor," the ratio of maximum power output to the product of V_{oc} and I_{sc} (the voltage and current for maximum output are known as V_m and I_m, respectively). These three parameters determine the efficiency and the circuit conditions to be used with the cell or an array of such cells. For satellite applications, a fourth parameter is of importance, the radiation damage coefficient K_L for various particles and energies. K_L describes the way in which the minority carrier diffusion length degrades as a function of radiation fluence.

The open circuit voltage of a p-n junction solar cell is directly related to the bandgap of the semiconductor through the energy barrier height at the junction; it is often written as a function of the short circuit photocurrent, the dark current I_0 of the junction, and the junction "perfection" factor A_0:

$$V_{oc} = A_0(kT/q) \ln[(I_{sc}/I_0)+1].\tag{1}$$

For a "perfect" junction, A_0 is equal to 1 and V_{oc} attains its highest value, while for larger values of A_0, I_0 is larger in such a way that V_{oc} is reduced. The logarithmic nature of the relation (1) causes V_{oc} to effectively saturate as a function of light intensity. The dark current I_0 is mainly determined by the bandgap of the material and the temperature; I_0 decreases and V_{oc} consequently increases with increasing bandgap or decreasing temperature.

The short circuit current I_{sc} is determined by the spectrum of the light source and the spectral response (electron-hole pairs collected per incident photon) of the cell. The spectral response in turn depends on the optical absorption coefficient α, the junction depth x_j, the width of the depletion region W, the lifetimes and mobilities on both sides of the junction, the presence or absence of electric fields on both sides of the junction, and the surface recombination velocity S. The energy contained in sunlight is distributed over a wide range of wavelengths, and efficient conversion requires a wide spectral response. The bandgap dependence enters through the absorption coefficient; generally speaking, wider bandgap materials absorb less sunlight and have smaller short circuit currents than narrow bandgap materials.

The fill factor (FF) is determined by the magnitude of the open circuit voltage, the value of A_0, and the series and shunt resistances R_s and R_{sh} (the internal resistances in series and in parallel with the p-n junction). The higher the V_{oc} and R_{sh}, and the lower the A_0 and R_s, the larger the FF will be.

The radiation damage coefficient K_L is a function of the type of particle and its energy, the lifetime of the minority carrier, the particular dopant in the base and its concentration, and the temperature both during and after irradiation.

This book is intended as a general review of the principles and technology of solar cells. Chapter 2 will deal with the physics underlying carrier generation and collection and determination of the short circuit current I_{sc}. Chapter 3 will cover the electrical characteristics of the p-n junction and voltage generation. Chapter 4 will deal with theoretical and experimental efficiencies and Chapter 5, with the effects of thickness. Chapter 6 will describe other types of solar cells, including Schottky barriers, heterojunctions, and vertical multijunctions. In Chapter 7, the effects of high energy particle radiation will be discussed, and in Chapter 8, the effects of temperature and intensity will be described. Chapter 9 will cover the technology of fabrication, diffusion, contacting, and applying antireflective coatings to the cells.

Since there are so many factors that affect solar cell
behavior, it is impossible to include all of them at the same
time and still perform numerical calculations that describe
general cases. For this reason, uniformly doped top and base
regions with good lifetimes are assumed as a starting point,
and short circuit currents, open circuit voltages, and efficiencies are calculated for this structure under the idealized
conditions of no series or shunt resistance losses, and no
reflection or contact area losses. The trends introduced by
decreasing values of lifetime, decreasing junction depths,
the presence of electric drift fields, and varying surface
recombination velocities are then described, and the effects
of the various loss mechanisms are discussed. In this way,
the technology-oriented features of solar cells can be separated from the inherent features, and a clearer understanding
of these devices can be obtained.

CHAPTER 2

Carrier Collection, Spectral
Response, and Photocurrent

A. Absorption and Lifetime

The energy band diagrams of illuminated p-n junction
solar cells for both short circuit conditions and when a load
is placed across the terminals are shown in Fig. 3. The
doping has been taken as uniform, and there are no built-in
electric fields outside of the space charge region (a slight
band bending at the surface due to surface states is indicated). When photons are incident with energy greater than
the bandgap, absorption of the photons can take place and
electrons can be raised in energy from the valence band to the
conduction band, creating hole-electron pairs. If the excess
minority carriers (holes on the n-side and electrons on the
p-side of the junction) are able to diffuse to the edges of
the space charge region before they recombine, they are "swept"
across the junction, giving rise to a photocurrent, photovoltage, and power into the load. The polarity of the output
voltage is the same as the "forward bias" direction of the
device, but the photocurrent is opposite in direction to the
forward bias current through the device in the dark.

The ability of a material to absorb light of a given
wavelength is measured quantitatively by the absorption coefficient α, measured in units of reciprocal distance. Light
incident at the surface falls off in intensity by a factor of
$1/e$ for each $1/\alpha$ distance into the material. As a general
rule, the larger the bandgap, the smaller the value of α is
for a given wavelength, but the absorption coefficient also
depends on the densities of states in the conduction and
valence bands and on the directness or indirectness of the
bandgap. Figure 4 shows the absorption coefficients of Si,
Ge, and GaAs in the range of 0.6-5.0 eV [12-16]. The Si coefficient rises very gradually due to the indirect bandgap
and consequently much of the absorption and carrier generation
occurs well below the Si surface (tens of microns). The GaAs

8

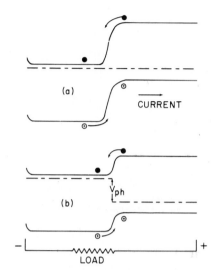

CURRENT

(a)

(b)

V_{ph}

LOAD

FIG. 3. An N/P junction
solar cell under illumination:
(a) short circuit conditions;
(b) when a load is added.

coefficient, on the other hand, rises very steeply at the
band edge, then increases gradually due to the low densities
of states. Most of the absorption and carrier generation in
GaAs occurs within 2 μm of the surface.

The collection of photogenerated carriers by movement
across the p-n junction is in competition with the loss of
these carriers by bulk and surface recombination before they
can be collected. Bulk recombination can occur by direct
mutual annihilation of a free electron and free hole or by
annihilation through an intermediate recombination center;
the intermediate recombination is usually the dominant mechanism. If there are N_r recombination centers located at an
energy E_r and having capture cross sections σ_n, σ_p for an
electron when empty and a hole when filled, respectively,
then the hole lifetime on the n side of the junction can be
described by [17]

$$\tau_p = (1/\sigma_p v_{th} N_r) \left[\left(1 + (N_c/n_{n0}) \ \exp\left[-(E_c-E_r)/kT\right]\right) \right.$$

$$\left. + (\sigma_p/\sigma_n) (N_v/n_{n0}) \ \exp\left[-(E_r-E_v)/kT\right]\right], \qquad (2)$$

where n_{n0} is the free electron concentration on the n-type
side and is essentially equal to the doping level, v_{th} is the
thermal velocity, k is Boltzmann's constant, and E_c and E_v are
the conduction and valence band edges, respectively. An analogous equation can be obtained for the electron lifetime in
p-type material. The actual lifetimes are probably determined

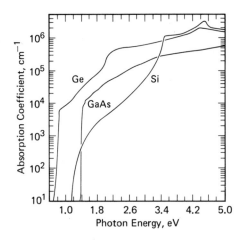

FIG. 4. Intrinsic absorption coefficients of Si, Ge, and GaAs.

by a multiplicity of such recombination centers at a number of energy levels, but qualitatively, this equation indicates that the lifetime will decrease with increasing doping level and saturate at a value equal to

$$\tau_{p0} = (\sigma_p v_{th} N_r)^{-1}. \tag{3}$$

Experimentally, the lifetime in Si and GaAs does decrease with increasing doping level and does saturate under some conditions [18], but as a rule there is no unique lifetime at a given doping level; the lifetime in bulk Si, for example, depends on the method of growth of the crystal (crucible grown or float-zoned), the duration and temperature of an annealing step (if any), and the rate at which the crystal is cooled [18-20]. For a finished device, the lifetime depends on the surface treatment during fabrication, the diffusion tempera- ture, the rate of cooling, and the presence or absence of annealing steps [18-20]. Table 1 lists some of the lifetimes observed in bulk Si under various conditions. The electron lifetime in as-grown crystals is higher in float-zone than in crucible-grown material, and it can be considerably in- creased by annealing at around 450°C [19,20], as shown in Fig. 5. Changes in lifetime produced by annealing are revers- ible up to about 600°C. For temperatures between 600 and 1250°C or so, an irreversible degradation in lifetime takes place [18-20], but above 1250°C, it appears that the starting lifetime (before annealing) can be at least partly restored [18].

TABLE 1
Lifetimes in Bulk Si, 300°K.
Boron-Doped Unless Otherwise Noted

ρ (ohm-cm)	Type	τ_n (10^{-6} sec)	Conditions	Ref.
10	p;FZ	55	As grown	19
		275	+ Annealed 450°C	
		10	+ Annealed 700°C	
1	p;CG	10-20	As grown	19
		20-120	+ Annealed 450°C	
1	p;FZ	110	As grown	20
		420	+ Annealed 450°C	
		15	+ Annealed 700°C	
1	p;CG	35	As grown	20
		210	+ Annealed 450°C	
		20	+ Annealed 700°C	
10	n	$\tau_p = 700$	Li-doped, as grown	21,22
1	n	$\tau_p = 200$	Li-doped, as grown	21,22
0.1	n	$\tau_p = 50$	Li-doped, as grown	21,22

Since the fabrication of solar cells involves several high temperature steps equivalent to high temperature annealing, it might be expected that the lifetimes measured in finished devices would be lower than those measured in the bulk. Table 2 shows the lifetimes measured in diffused devices of several different base resistivities; the lifetimes tend to be lower than those in the bulk Si by a factor of 2 to 5.

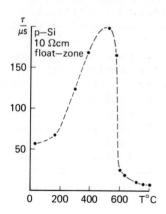

FIG. 5. The effect on electron lifetime of annealing for 1 hr at various temperatures. (After Graff et al. [19]; courtesy of the Electrochemical Society.)

TABLE 2
Lifetimes in Si-Diffused Devices, 300°K.
Boron or Arsenic Doped

ρ Base (ohm-cm)	Type	τ_n (10^{-6} sec)	τ_p (10^{-6} sec)	Conditions	Ref.
10	n and p,	8.5	10.5	As grown	18
1	CG	2.3	2.8	As grown	18
0.1	CG	1.2	0.5	As grown	18
1	p;CG	4-10		After diffusion	20
		9-20		After annealing at 200°C	
10	p	7.1			23
2	p	4.4			23
10	p	10-25			24
1	p	3-12			24

In the diffused top region of the cell, lattice damage, very
high doping levels, and unwanted impurity incorporation into
the material tend to introduce high recombination center densities; estimates have been made that the lifetime can be as
low as 100 psec within the first 1000 Å or so from the surface
(the so-called "dead region") [4] and only a few nanoseconds
in the remainder of the top region [4,24]. The very high
doping level (up to 5 at. %) at the surface of the diffused
region can actually lead to an effective reduction of the
bandgap at the surface, as well as very short lifetimes.
Both of these effects contribute to the probability of a dead
region being present near the surface.

The variation in lifetime with annealing temperature,
rate of cooling, surface preparation, and the method of crystal
growth has been attributed speculatively to the behavior of
oxygen in Si [19,20]. Annealing at low temperatures, it has
been suggested [19], produces oxygen-donor complexes and
reduces the density of recombination centers. At around 600°C,
the complexes transform to other configurations and the density of recombination centers increases. Above 600°C, irreversible precipitation of oxygen begins to take place, which
lowers the lifetime irreversibly until temperatures of 1250-
1350°C are reached. The higher lifetimes observed in float-
zone Si compared to crucible grown material are attributed to
the normally lower oxygen content of float zone material, as
discussed in Chapter 9.

The lifetimes of direct gap materials such as GaAs tend
to be much shorter than those of Si or Ge. Table 3 lists the
hole and electron lifetimes and the diffusion lengths (defined
as the distance a free charge carrier can diffuse in one lifetime period, and mathematically given by L = $\sqrt{D\tau}$) of GaAs
for various doping levels and for several types of dopant.
For p-type GaAs, the lifetime is higher in liquid phase epitaxy (LPE) material than in crucible or vapor grown material,
and higher for Ge doping than for Zn. This appears to be due
in part to the "gettering" behavior of the liquid Ga used in
the LPE technique, whereby impurities are removed from a GaAs
crystal in contact with the melt and retained in the Ga, and
in part to the better match of the Ge atom to the GaAs lattice
compared to the Zn atom, producing less strain with Ge doping
than with Zn. For n-type material, there is some evidence
that Si doping yields better lifetimes than Sn, S, Se, or Te,
and the Si atom does match the GaAs lattice better than these
others, but the evidence is not as strong as in the p-type
case.

In addition to recombination in the bulk, a loss of
photogenerated minority carriers also takes place at the surfaces of the material due to the presence of surface states
which arise from "dangling bonds," chemical residues, metal
precipitates, native oxides, and the like. The rate at which
carriers are lost at a surface is described by the surface
recombination velocity S; the minority carrier current density
toward the surface is given by

$$J_{surface} = qS_p(p_n-p_{n0}) \qquad \text{n-type material,}$$

$$= qS_n(n_p-n_{p0}) \qquad \text{p-type material.}$$

(4)

The recombination velocity at the illuminated surface is of
critical importance, since the number of carriers generated
for a given wavelength of light is highest at this surface
and decreases exponentially with distance into the cell. The
high value of S at this front surface, together with the low
bulk lifetime that usually occurs in the diffused top region,
necessitates shallow junction depths (0.5 μm or less) in order
to prevent a serious loss of carriers. The recombination
velocity at the back of the cell is not as critical, but its
importance increases as cells are made thinner, particularly
for lightly doped base regions.

TABLE 3

Lifetimes and Diffusion Lengths, GaAs, 300°K

Doping (cm^{-3})	Dopant	τ_n (10^{-8} sec)	L_n (10^{-4} cm)	τ_p (10^{-8} sec)	L_p (10^{-4} cm)	Ref.
2×10^{17}	?	0.35	6–6.3	–	–	25
2×10^{18}	?	0.092	1.9–3.3	–	–	25
5×10^{18}	Zn(LPE)[a]	0.63	6	–	–	26
>3×10^{18}	Zn(LPE)	>0.65	6–7	–	–	27
2×10^{18}	Zn(Boat)	0.217	4	–	–	28
1×10^{19}	Zn(Boat)	0.057	1.6	–	–	29
1×10^{18}	Ge(LPE)	5.88	23	–	–	30
5×10^{18}	Ge(LPE)	5.73	18	–	–	30
2×10^{18}	Ge(LPE)	1.50	10.5	–	–	31
1×10^{19}	Ge(LPE)	0.67	5.5	–	–	31
1×10^{17}	Ge,Sn	0.49	7.5	0.79	2.2	32
1×10^{18}	Ge,Sn	0.40	6	0.77	1.9	32
2×10^{18}	Sn	–	–	0.36	1.2	32
5×10^{18}	Ge	0.071	2	–	–	32
2×10^{17}	(Si ?)	–	–	1.9	3.1–3.5	33

[a] LPE = liquid-phase epitaxy.

Surface recombination at the front is even more important for direct bandgap materials like GaAs, where most carriers are generated close to the surface, than for indirect ones like Si where many carriers are generated deep in the material due to the low absorption coefficient at long wavelengths (Fig. 4). On the other hand, recombination at the back surface is more important for the indirect gap, long lifetime materials such as Si than for direct gap, low lifetime materials like GaAs. The recombination velocity at the front of most Si and GaAs solar cells is in the range 10^5-10^6 cm/sec, although etching the surfaces of bulk Si crystals has been known to reduce S to around 10^2 cm/sec in some cases.

For polycrystalline devices, such as thin film Si and GaAs solar cells, the recombination velocity at the grain boundaries is important, and if the grain size is much less than the diffusion length in a single crystal of the same doping level, the effective lifetime and diffusion length in the polycrystalline film are greatly reduced below their single crystal values.

B. Derivation of the Photocurrent for Monochromatic Light

The use of analytical tools to predict the behavior of solar cells, and in particular to predict the effect of variables such as doping level and junction depth, has proven to be very valuable in the past, even though the assumptions needed to obtain analytical expressions are violated to a degree in actual devices. The lifetime, mobility, and doping level in the base of most solar cells are reasonably constant, but are functions of position in the top region if this region is produced by diffusion. Numerical analyses can be made which take these variations into account provided that these functions are known. However, analytical expressions obtained by assuming average, constant values for these parameters and by assuming constant electric fields serve as useful first approximations in predicting the expected behavior, and are far less time consuming to obtain and interpret than the numerical results. Numerical methods will be particularly valuable when very strong variations in device parameters as a function of position are expected and when sunlight concentration by a factor of 20 or more is used so that "low injection level" conditions no longer apply.

When light of wavelength λ is incident on the surface of a semiconductor, the generation rate of hole-electron pairs as a function of distance x from the surface is

$$G(\lambda) = \alpha(\lambda)F(\lambda)[1-R(\lambda)] \exp(-\alpha(\lambda)x), \qquad (5)$$

where $F(\lambda)$ is the number of incident photons per cm^2 per sec per unit bandwidth and R is the number reflected from the surface. The photocurrent that these carriers produce and the spectral response (the number of carriers collected per incident photon at each wavelength) can be determined for low injection level conditions using the minority carrier continuity equations

$$(1/q)(dJ_p/dx)-G_p+(p_n-p_{n0})/\tau_p = 0 \qquad (6)$$

for holes in n-type material, and

$$(1/q)(dJ_n/dx)+G_n-(n_p-n_{p0})/\tau_n = 0 \qquad (7)$$

for electrons in p-type material. The hole and electron currents are

$$J_p = q\mu_p p_n E-qD_p(dp_n/dx), \qquad (8)$$

$$J_n = q\mu_n n_p E+qD_n(dn_p/dx), \qquad (9)$$

respectively, where E is the electric field, p_n and n_p are the photogenerated minority carrier densities, and p_{n0}, n_{p0} are the minority carrier densities in equilibrium in the dark.

P/N junction solar cells can be represented by one of several different physical models, depending on how they are made. In the simplest model, both sides of the junction are taken to be uniform in doping, mobility, and lifetime; this can be used to describe devices with grown top regions and as a first approximation to devices with diffused top regions. In the second model, electric fields exist in the base and/or top region as a result of doping nonuniformities, but the mobility and lifetime are still taken as constant in order to obtain an analytical result. If the mobility and lifetime are allowed to vary, numerical methods can be used to obtain the spectral response and short circuit current. In the third model, the base is divided into two sections with distinctly different properties. This applies, for example, to the "back surface field" cell [5] where a p^+ region is diffused into the back of the p-type base to enhance both the short circuit current and open circuit voltage of the cell.

In those instances where a "dead layer" is believed to exist adjacent to the surface of the device due to stress introduced by the junction diffusion, the n-type diffused top region of the cell can also be divided into two sections with different lifetimes and mobilities.[a]
In each of these models, Eqs. (5)-(9) can be applied to the appropriate regions and proper boundary conditions can be used to obtain the spectral response and the photocurrent.

1. UNIFORM N/P JUNCTIONS

If the two sides of the junction are uniform in doping, then there are no electric fields outside of the depletion region. This model applies to an epitaxially grown junction or as a first approximation to a diffused junction. For the case of an N/P device, where the base is p-type and the top side is n-type, Eqs. (6) and (8) can be combined to yield

$$D_p \frac{d^2(p_n-p_{n0})}{dx^2} + \alpha F(1-R)\exp(-\alpha x) - \frac{(p_n-p_{n0})}{\tau_p} = 0 \qquad (10)$$

for the top side of the junction. The general solution to this is

$$(p_n-p_{n0}) = A\cosh\left(\frac{x}{L_p}\right) + B\sinh\left(\frac{x}{L_p}\right) - \frac{\alpha F(1-R)\tau_p}{(\alpha^2 L_p^2 - 1)}\exp(-\alpha x), \qquad (11)$$

where L_p is the diffusion length, $L_p = (D_p\tau_p)^{1/2}$. For a single crystal device, there are two boundary conditions; at the surface, recombination takes place:

$$D_p \frac{d(p_n-p_{n0})}{dx} = S_p(p_n-p_{n0}) \qquad [x = 0] \qquad (12)$$

[a]Physically, a top region with a narrow dead section near the surface and a wider section of higher lifetime near the junction edge is essentially equivalent to a uniform top region with a high recombination velocity at its surface. Therefore, a "dead layer" in this book is modeled as a uniform top region with a 3 nsec lifetime and a surface recombination velocity of 10^6 cm/sec or higher.

while at the junction edge, the excess carrier density is reduced to zero by the electric field in the depletion region

$$p_n - p_{n0} = 0 \quad [x = x_j].\tag{13}$$

Using these boundary conditions in (11), the hole density is found to be

$$(p_n - p_{n0}) = \left[\alpha F(1-R)\tau_p / (\alpha^2 L_p^2 - 1)\right]$$

$$\times \left[\frac{\left(\dfrac{S_p L_p}{D_p} + \alpha L_p\right)\sinh\dfrac{x_j - x}{L_p} + \exp(-\alpha x_j)\left(\dfrac{S_p L_p}{D_p}\sinh\dfrac{x}{L_p} + \cosh\dfrac{x}{L_p}\right)}{\dfrac{S_p L_p}{D_p}\sinh\dfrac{x_j}{L_p} + \cosh\dfrac{x_j}{L_p}} - \exp(-\alpha x)\right]\tag{14}$$

and the resulting hole photocurrent density per unit band-width at the junction edge is

$$J_p = \left[\frac{qF(1-R)\alpha L_p}{(\alpha^2 L_p^2 - 1)}\right]$$

$$\times \left[\frac{\left(\dfrac{S_p L_p}{D_p} + \alpha L_p\right) - \exp(-\alpha x_j)\left(\dfrac{S_p L_p}{D_p}\cosh\dfrac{x_j}{L_p} + \sinh\dfrac{x_j}{L_p}\right)}{\dfrac{S_p L_p}{D_p}\sinh\dfrac{x_j}{L_p} + \cosh\dfrac{x_j}{L_p}} - \alpha L_p \exp(-\alpha x_j)\right].\tag{15}$$

This is the photocurrent that would be collected from the top side of a N/P junction solar cell at a given wavelength, as-suming this region to be uniform in lifetime, mobility, and doping level.

To find the electron current collected from the base of the cell, (7) and (9) are used, making the same approximation as before that the base is uniform in its electrical proper-ties. The boundary conditions are:

$$(n_p-n_{p0}) = 0 \qquad [x = x_j+W], \tag{16}$$

$$S_n(n_p-n_{p0}) = -D_n[d(n_p-n_{p0})/dx] \qquad [x = H], \tag{17}$$

where W is the width of the depletion region and H is the width of the entire cell. Equation (16) states that the excess minority carrier density is reduced to zero at the edge of the depletion region, while (17) states that surface recombination takes place at the back of the cell. (If the back is covered with an Ohmic contact, a perfect "sink" for the minority carriers exists and S_n can be taken as infinite.)

Using these boundary conditions, the electron distribution in a uniform p-type base is

$$(n_p-n_{p0}) = \frac{\alpha F(1-R)\tau_n}{(\alpha^2 L_n^2-1)} \exp[-\alpha(x_j+W)]\left[\cosh\frac{x-x_j-W}{L_n} -\exp[-\alpha(x-x_j-W)]\right.$$

$$\left. -\frac{\left(\dfrac{S_n L_n}{D_n}\right)\left[\cosh\dfrac{H'}{L_n}-\exp(-\alpha H')\right]+\sinh\dfrac{H'}{L_n}+\alpha L_n \exp(-\alpha H')}{(S_n L_n/D_n)\ \sinh(H'/L_n)+\cosh(H'/L_n)} \sinh\frac{x-x_j-W}{L_n}\right] \tag{18}$$

and the photocurrent per unit bandwidth due to electrons collected at the junction edge is

$$J_n = \frac{qF(1-R)\alpha L_n}{(\alpha^2 L_n^2-1)} \exp[-\alpha(x_j+W)]$$

$$\times\left[\alpha L_n - \frac{\dfrac{S_n L_n}{D_n}\left(\cosh\dfrac{H'}{L_n}-\exp(-\alpha H')\right)+\sinh\dfrac{H'}{L_n}+\alpha L_n \exp(-\alpha H')}{\dfrac{S_n L_n}{D_n}\sinh\dfrac{H'}{L_n}+\cosh\dfrac{H'}{L_n}}\right] \tag{19}$$

where H' is the total cell thickness minus the junction depth and depletion width, $H' = H-(x_j+W)$.

Some photocurrent collection takes place from the depletion region as well. The electric field in this region can be considered high enough that photogenerated carriers are accelerated out of the depletion region before they can recombine, so that the photocurrent per unit bandwidth is equal simply to the number of photons absorbed

$$J_{dr} = qF(1-R) \ \exp(-\alpha x_j)[1-\exp(-\alpha W)]. \tag{20}$$

The total short circuit photocurrent at a given wavelength is then the sum of (15), (19), and (20),[b] and the spectral response is equal to this sum divided by $qF(1-R)$ (internal response) or qF (externally observed response).

All of these equations from (12) to (19) can be transformed from their present form for N/P cells to equivalents for P/N cells by interchanging L_n, D_n, τ_n, and S_n with L_p, D_p, τ_p, and S_p, respectively.

2. CONSTANT ELECTRIC FIELDS

Losses due to surface and bulk recombination can be reduced by decreasing the junction depth and surface recombination velocity and by increasing the minority carrier diffusion length, particularly in the base. Another way that losses can be reduced is to provide electric fields in one or both regions of the cell to aid in moving photogenerated carriers toward the junction. (In theory, diffusion processes provide electric fields automatically by establishing concentration gradients of donors or acceptors. In reality, the dependence of the diffusion coefficient on the impurity concentration can almost eliminate the gradient and the hoped-for drift field [34-36].) When such a field is present, the energy band edges are sloped, rather than flat as in Fig. 3, and the field at any point is given by the slope

$$E = (1/q)(dE_c/dx) = (1/q)(dE_v/dx) = (D/\mu)(1/N)(dN/dx) \tag{21}$$

where N is the ionized impurity concentration. The field in the diffused region is largest at the edge of the junction and smallest at the surface of the device; it would be better for the purposes of overcoming surface recombination if it were the other way around.

Using (6) and (8), the continuity equation for holes in the diffused n-type region becomes

$$(d/dx)[D_p(dp_n/dx)]-(d/dx)(\mu_p p_n E)$$

$$+\alpha F(1-R) \ \exp(-\alpha x)-[(p_n-p_{no})/\tau_p] = 0 \tag{22}$$

[b]These equations are valid for all wavelengths except the special case where $\alpha L = 1$, and they have the correct limit as $\alpha L \to 1$. For discussions and derivations of this special case, reference is made to Wolf [34].

with the boundary conditions

$$D_p(dp_n/dx) - \mu_p p_n E = S_p(p_n - p_{n0}) \qquad [x = 0], \tag{23}$$

$$(p_n - p_{n0}) = 0 \qquad [x = x_j]. \tag{24}$$

Wolf [34], Ellis and Moss [37], and others have solved these relationships for the special case where the electric field, mobility, and lifetime are taken to be constant across the diffused region. The photocurrent from the diffused region is then [34]

$$
J_p = \frac{qF(1-R)\alpha L_{pp}}{(\alpha+E_{pp})^2 L_{pp}^2 - 1} \left[\frac{(\alpha+E_{pp})L_{pp}\ \exp(E_{pp}x_j) - \exp(x_j/L_{pp})\exp(-\alpha x_j)}{\left(\dfrac{S_p L_{pp}}{D_p} + E_{pp}L_{pp}\right)\sinh\dfrac{x_j}{L_{pp}} + \cosh\dfrac{x_j}{L_{pp}}} \right.
$$

$$
+ \frac{\left(\dfrac{S_p L_{pp}}{D_p} + E_{pp}L_{pp}\right)\left(\exp(E_{pp}x_j) - \exp(x_j/L_{pp})\exp(-\alpha x_j)\right)}{\left(\dfrac{S_p L_{pp}}{D_p} + E_{pp}L_{pp}\right)\sinh\dfrac{x_j}{L_{pp}} + \cosh\dfrac{x_j}{L_{pp}}}
$$

$$
\left. -\exp(-\alpha x_j)\left[(\alpha+E_{pp})L_{pp} - 1\right] \right] \tag{25}
$$

where E_{pp} is the normalized electric field in the n-type diffused region

$$E_{pp} = qE/2kT \tag{26}$$

and L_{pp} is an effective diffusion length

$$L_{pp}^{-1} = \sqrt{E_{pp}^2 + (1/L_p^2)} \tag{27}$$

When the electric field, lifetime, and mobility cannot be approximated as constant throughout the diffused region, the continuity Eq. (22) becomes far more complex and an analytical solution cannot be obtained. Numerical methods must then be used as has been done by Tsaur et al. [38] for GaAs solar cells, or Fossom [39] for Si solar cells.

The equations above are valid provided that the carriers accelerated by the field do not reach their saturation drift velocity, which occurs at junction depths of around 1000 Å.

In addition to a drift field in the top region, it is possible to obtain an electric field in the base as well by providing a concentration gradient there, with the lowest doping concentration at the junction edge and increasingly higher concentrations toward the back contact. This electric field aids the collection of carriers generated at low photon energies and is capable of improving the resistance of the cell to radiation degradation [34,37,40-42]. The photocurrent in the p-type base can be calculated from (7) and (9)

$$(d/dx)\,[D_n(dn_p/dx)]+(d/dx)\,(\mu_n n_p E)$$

$$+\ F(1-R)\ \exp(-\alpha x)\ \exp[\alpha(x_j+W)]-[(n_p-n_{p0})/\tau_n] = 0 \qquad (28)$$

with the boundary conditions

$$(n_p-n_{p0}) = 0 \qquad [x = x_j+W], \qquad\qquad\qquad (29)$$

$$D_n(dn_p/dx)+\mu_n n_p E = -S_n(n_p-n_{p0}) \qquad\qquad [x = H] \qquad (30)$$

which are analogous to (16) and (17). If the approximations are made that the field, lifetime, and mobility are constant in the base, the photocurrent at a given wavelength derived from these relationships is [34]

$$
J_n = \frac{q\alpha F(1-R)L_{nn}\ \exp(-E_{nn}x_j)\ \exp(-\alpha W)}{(\alpha-E_{nn})^2\ L_{nn}^2-1}
$$

$$
\times \left[\,[\,(\alpha-E_{nn})L_{nn}-1]\ \exp[-(\alpha-E_{nn})x_j]+\exp[-(\alpha-E_{nn})(H-W)]\,\right]
$$

$$
\times \left[\frac{\left(\begin{array}{l}\exp(-H'/L_{nn})\ \exp[(\alpha-E_{nn})H']-(\alpha-E_{nn})L_{nn}\\[4pt] \qquad -(E_{nn}+S_n/D_n)L_{nn}[\exp(-H'/L_{nn})\exp((\alpha-E_{nn})H')-1]\end{array}\right)}{(E_{nn}+S_n/D_n)L_{nn}\ \sinh(H'/L_{nn})+\cosh(H'/L_{nn})} \right]
$$

$$(31)$$

where Wolf's [34] equation has been modified slightly to account for the finite width of the depletion region, and E_{nn} and L_{nn} are defined in the same way as (26) and (27).

Without the assumptions of constant material parameters as a function of position in the base, an analytical result for the photocurrent cannot be obtained, and numerical computations must be used [40-42]. None of the numerical calculations made so far have been completely satisfactory, since all of them involve approximations and assumptions of various types. The most rigorous calculation to date has been that

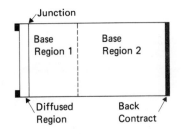

FIG. 6. Diagram of a solar cell in which the base can be divided into two regions with different doping levels, mobilities, lifetimes, and electric fields.

of Van Overstraeten and Nuyts [42]. They divide the base into a first region near the junction where the field, lifetime, and mobility are all variables, and a second region extending to the back Ohmic contact in which the lifetime and mobility are constant and the field is zero (Fig. 6). Using numerical integration, they calculated the photocurrent that would be obtained under various conditions in Si devices and showed that the photocurrent is most strongly affected by the first 10 or 20 μm of the base adjacent to the junction edge; the impurity concentration should be small and the mobility, lifetime, and electric field should be high in this first 20 μm to obtain the highest collection efficiency.

3. BACK SURFACE FIELD, HIGH-LOW JUNCTION

A dramatic improvement in the output voltage of Si solar cells has been noted in the last few years with the advent of the "back surface field" (BSF) cell [5,43,44]. In this device, the front (junction) part of the cell is made in the normal way, but the back of the cell, instead of containing just a metallic Ohmic contact to the moderately high resistivity base, has a very heavily doped region adjacent to the contact. In Fig. 6 for example, base region 1 represents the normal 1 to 10 ohm-cm portion while region 2 represents the more heavily doped layer adjacent to the contact. Region 2 is typically only a micron or two wide if made by diffusion or alloying, but in some schemes region 1 is made narrow (10 μm) and region 2, wide by growing a lightly doped epitaxial layer on a heavily doped substrate.

The advantages of the extra region can be seen with the help of the band diagram of Fig. 7. The potential energy barrier ψ_p between the two base regions tends to "confine" minority carriers in the more lightly doped region, away from the Ohmic contact at the back with its infinite surface recombination velocity. If W_p is comparable to or less than the

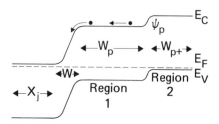

FIG. 7.   Energy band diagram of a BSF (blocking back contact) device.

diffusion length L_n in region 1, then some of the electrons that would have been lost at the back surface cross the p-n junction boundary instead, enhancing the short circuit current. To a first approximation, the BSF cell can be modelled as a normal cell of width (x_j+W+W_p) having a very small recombina- tion velocity at the back [Eqs. (19) or (31) with $S_n = 0$] provided $W_p \gg W_{p+}$.

The open circuit voltage of N on P BSF cells with 10 ohm- cm base resistivity is around 10% higher than conventional cells of the same type (0.6 V compared to 0.55 V), probably due to a combination of three factors: the increased short circuit current [see Eq. (1)], a decrease in I_0 (the diode "leakage" current) due to reduced recombination at the *back* surface of electrons injected from the n^+ top region into the base, and a modulation of the barrier ψ_p by the change in minority carrier densities at the high-low junction edge, i.e., when the cell is open circuited, a portion of the barrier ψ_p might appear at the output terminals in addition to the volt- age from the p-n junction.

C. Spectral Response

The photocurrent collected at each wavelength relative to the number of photons incident on the surface at that wave- length determines the spectral response of the device (some- times known as the quantum efficiency or collection efficiency at each wavelength). The "internal" spectral response is defined as the number of electron-hole pairs collected under short circuit conditions relative to the number of photons *entering* the material

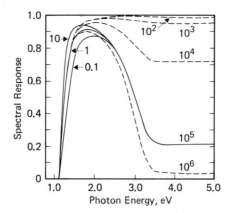

FIG. 8.   *Computed internal spectral responses of Si N/P cells with uniformly doped regions. The solid lines are for 10, 1, and 0.1 ohm-cm base resistivities and $S_{front} = 10^5$ cm/sec. The dotted lines are for 1 ohm-cm bases with varying S_{front}. Other parameters, H = 450 μm, x_j = 0.5 μm, S_{back} = ∞. No drift field.*

$$SR(\lambda) = \frac{J_p(\lambda)}{qF(\lambda)(1-R(\lambda))} + \frac{J_n(\lambda)}{qF(\lambda)(1-R(\lambda))} + \frac{J_{dr}(\lambda)}{qF(\lambda)(1-R(\lambda))} \qquad (32)$$

while the "external" response is just the internal one modified by reflection of light from the surface of the device

$$SR(\lambda)_{ext} = SR(\lambda)[1-R(\lambda)]. \qquad (33)$$

The reflection of light from the surface as a function of wavelength enters into the experimentally observed spectral response, but in general the technology of antireflective coatings on solar cells has been developed to such a high degree that the reflection and its variation with wavelength can be ignored to a first approximation when comparing measured spectral response curves to predicted ones.

1.   CALCULATED RESPONSES

The internal spectral responses of Si N/P cells for the simplest case of uniformly doped top and base regions are shown in Fig. 8, as calculated from (15), (19), (20), and (32) using the device parameters listed in Table 4.   The solid lines demonstrate the effect of varying the resistivity

TABLE 4

Solar Cell Parameters for Si, 300°K

N on P cells. $N_d = 5\times10^{19}$, $D_p = 1.295$, $\tau_p = 0.4\times10^{-6}$

ρ_{Base}	N_a	μ_n	D_n	τ_n	L_n	W(0 bias)	V_d
(Ω-cm)	(cm^{-3})	(cm^2/V·sec)	(cm^2/sec)	(sec)	(10^{-4} cm)	(10^{-4} cm)	(V)
10	1.25×10^{15}	1390	36	15×10^{-6}	232	0.93	0.867
1	1.5×10^{16}	1040	27	10×10^{-6}	164	0.28	0.930
0.1	5×10^{17}	420	10.9	2.5×10^{-6}	52.2	0.05	1.022

P on N cells. $N_a = 5\times10^{19}$, $D_n = 2.15$, $\tau_n = 1.1\times10^{-6}$

ρ_{Base}	N_d	μ_p	D_p	τ_p	L_p	W(0 bias)	V_d
(Ω-cm)	(cm^{-3})	(cm^2/V·sec)	(cm^2/sec)	(sec)	(10^{-4} cm)	(10^{-4} cm)	(V)
10	4.5×10^{14}	580	15	15×10^{-6}	150	1.5	0.814
1	5.1×10^{15}	500	13	7.5×10^{-6}	98.5	0.47	0.877
0.1	8.5×10^{16}	350	9	1.5×10^{-6}	36.9	0.12	0.950

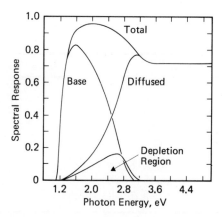

FIG. 9.   Computed internal spectral response of a Si N/P cell showing the individual contributions from each of the three regions.

of the base while keeping the surface recombination velocity and junction depth constant. Raising the doping level in the base lowers the lifetime and diffusion length in the base, increasing the loss of carriers generated deep in the material and degrading the low energy response. The dashed lines show what happens to the high energy response at various surface recombination velocities for constant base parameters. At high photon energies, all the carriers are generated near the surface because of the high absorption coefficients at these energies, and losses due to high S_p or poor lifetime in the diffused N-type region become critical. Above 3.5 eV where the spectral response derives entirely from the N surface region, the response (32) saturates at a value given by

$$SR = \frac{1+(S_p/\alpha D_p)}{(S_p L_p/D_p)\ \sinh(x_j/L_p)+\cosh(x_j/L_p)} . \qquad (34)$$

Since the Si absorption coefficient α is relatively constant at $1\text{-}2\times10^6$ cm^{-1} from 3.5 to 4.0 eV (Fig. 4), this relationship could be used to estimate S_p from a measured spectral response if the value of D_p is known in the diffused N region.

For low values of surface recombination velocity the response remains high and relatively flat over the whole spectral region. This is the type of response that should be obtained for "violet cells," Schottky barrier cells, and certain types of heterojunctions with thin, high bandgap transparent semiconductor layers on Si or GaAs substrates.

TABLE 5
Solar Cell Parameters for GaAs, 300°K. P on N

Top Region	Base Region
N_a = 2×10^{19} cm^{-3}	N_d = 2×10^{17} cm^{-3}
D_n = 32.4 cm^2/sec	D_p = 5.7 cm^2/sec
τ_n = 1×10^{-9} sec	τ_p = 1.58×10^{-8} sec
L_n = 1.8 µm	L_p = 3.0 µm

$$W = 0.09 \text{ µm}$$
$$V_d = 1.40 \text{ V}$$
$$n_i = 1.1×10^7 \text{ cm}^{-3}$$

In Fig. 9, the spectral response of the 1 ohm-cm N/P Si cell with S_p = 10^4 cm/sec as shown in Fig. 8 is divided into its three components, the base, the diffused top region, and the depletion region contributions. At low energies, most of the carriers are generated in the base because of the low absorption coefficients, but as the photon energy increases above 2.4 eV, the diffused side of the junction takes over. If the junction depth is made smaller than the 0.5 µm shown here, the contribution from the base increases slightly and the crossover moves to slightly higher energies, but more importantly, the contribution from the diffused side at high energies is enhanced because of reduced losses due to surface and bulk recombination. Most N/P cells made today have junction depths in the 0.3-0.5 µm range, while the "violet cell" has a junction depth of only 0.1-0.2 µm.

The contribution from the depletion region is considerable in the 2.0 to 2.9 eV range for the 1 ohm-cm device shown in Fig. 9. The depletion region contribution becomes greater at higher base resistivities and narrower junction depths, and less for lower resistivities and larger depths, but it never becomes as large as the diffused region component under any practical conditions because of the very high value of α above 3.2 eV (Fig. 4), which causes almost all the light at high energies to be absorbed in the first 1000 Å or so.

The calculated internal spectral responses of N/P GaAs cells with uniformly doped regions for several surface recombination velocities are shown in Fig. 10, using the device parameters of Table 5. GaAs is a direct bandgap material with a steep absorption edge. Virtually all the carriers generated by sunlight above 1.4 eV are generated in the first 3 µm from the surface, and 50% of all the carriers are generated within the first 1/2 µm. This makes the properties of the top side

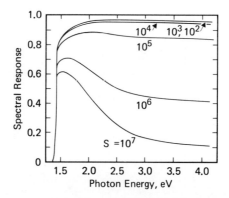

FIG. 10.   *Computed internal spectral responses of GaAs P/N solar cells with uniformly doped top and base regions for various front surface recombination velocities.   H = 300 μm, x_j = 0.5 μm, S_{back} = ∞.   Device parameters of Table 5.*

of the junction much more important than in silicon, and the base of the cell correspondingly less important. High values of front surface recombination velocity and low values of life- time and diffusion length in the top region cause a strong decrease in the spectral response with increasing photon energy as carriers are generated closer and closer to the surface. These high surface losses can be partly overcome by making the junction depth small, as in Si cells, or by establishing an aiding electric field at the surface.

To demonstrate the value that aiding electric fields can have in improving the carrier collection efficiency, the spec- tral response of an N/P Si cell was calculated from (20), (25), and (31) both with and without electric fields; the results are seen in Fig. 11. To illustrate the effect of the fields more strikingly, relatively poor cell conditions were adopted for the calculations. The lifetime in the diffused top region has been assumed to be low, corresponding to measured lifetimes in Si cells, and the surface recombination velocity has been assumed high. (For the uniform doping (zero field) case, the base resistivity has been taken as 1 ohm-cm. For the calcula- tion including electric fields, the base is assumed to have a doping level of a few times 10^{14} cm^{-3} at the junction edge with an increase to a few times 10^{17} several hundred microns away, resulting in an average field of 10 V/cm. The field in the diffused N-type region is taken as 4400 V/cm.)

The electric field in the base region enhances the response at low energies by drifting carriers toward the junction that might ordinarily be lost deep inside the Si. (The contribution

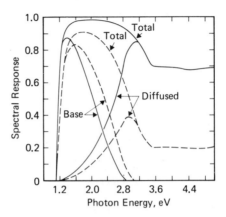

FIG. 11.   Computed internal spectral responses of Si N/P cells both with (solid) and without (dashed) electric fields.   (The contribution from the depletion region is not shown individually, but is included in the total.) $S_{front} = 10^5$ cm/sec, $\tau_p = 5\times10^{-9}$ sec, $H = 450$ μm, $x_j = 0.5$ μm, $S_{back} = \infty$.

from the base in the field case for energies above 1.6 eV can actually be a bit less than for uniform doping, however, because the depletion region width is greater for the lower doping at the junction edge in the field case compared to the zero field case, and some of the carriers that would have been collected from the base in the uniformly doped cell are generated and collected in the depletion region instead in the drift field cell. The sum of the contributions from the base and the depletion region is higher when the base drift field exists than for uniform doping.)

The most dramatic improvement due to electric fields is seen in the response from the diffused top region. The combination of poor lifetime and high recombination velocity at the front surface causes a strong decrease in the response with increasing photon energy when no field is present; the response is considerably better if an aiding electric field is present in the top region.

The same improvement in spectral response can be seen in Fig. 12 for a GaAs P/N cell with a field of 1280 V/cm in the 0.5 μm wide diffused top region. The recombination velocities on GaAs surfaces tend to lie in the 10^6-10^7 range, an order-of-magnitude or more greater than the values found on Si surfaces. Incorporating an aiding drift field and/or reducing the junction depth can be very beneficial in overcoming the high carrier losses experienced in the diffused top region.

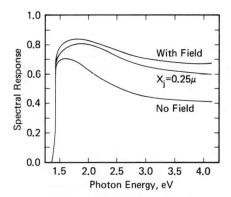

FIG. 12. *Computed internal spectral responses of GaAs P/N cells, showing the improvement obtained either by including a drift field in the top region or by reducing the junction depth.* *(x_j = 0.5 µm unless otherwise noted. No field in the base. Device parameters of Table 5. S = 10^6 cm/sec. S_{back} = ∞.)*

Another method for overcoming the effects of surface recombination and reducing losses due to bulk recombination in the diffused region is to grow a heavily doped, high band-gap semiconductive layer on the surface of the diffused region, choosing a material which closely matches the lattice proper-ties of the cell material. This method has been used with considerable success for GaAs solar cells [8,9,36,45], where $Ga_{1-x}Al_xAs$ is grown by liquid-phase epitaxy onto GaAs sub-strates. A schematic of the structure and the energy band diagram are shown in Fig. 13. The $Ga_{1-x}Al_xAs$ layer is trans-parent to most sunlight, and eliminates the surface states

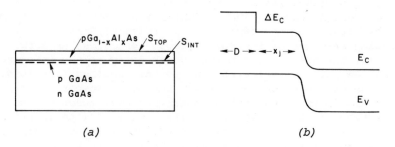

FIG. 13. *$Ga_{1-x}Al_xAs$-GaAs solar cell: (a) structure, (b) energy band diagram.*

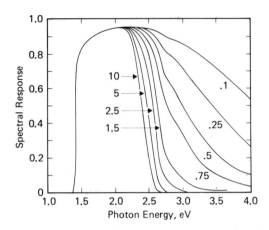

FIG. 14.   Computed internal spectral responses of $pGa_{1-x}Al_xAs$-pGaAs-nGaAs cells as a function of the $Ga_{1-x}Al_xAs$ thickness (in microns).   Aluminum composition is 0.86. $N_a = 2\times10^{19}$ cm^{-3}, $N_d = 2\times10^{17}$ cm^{-3}, $S_{top} = 10^6$ cm/sec, $S_{interface} = 10^4$ cm/sec. $L_{top} = 0.27$ µm, $L_{pGaAs} = 1.8$ µm, $L_{nGaAs} = 3.0$ µm, $x_j = 0.5$ µm, $H = 300$ µm, $S_{back} = \infty$.

and other imperfections on the GaAs P/N junction surface that
would ordinarily result in a high recombination velocity.   The
p-type layer also forms an Ohmic contact to the pGaAs region
and allows this region to be more lightly doped, which improves
the lifetime and lowers the bulk recombination but without
creating series resistance problems.   Photogenerated electrons
in the pGaAs region are prevented from entering the $Ga_{1-x}Al_xAs$
layer by the energy discontinuity ΔE_c in the conduction band
that arises from the difference in electron affinities of the
two materials.

Figure 14 shows the computed spectral response of a
$Ga_{1-x}Al_xAs$-GaAs solar cell as a function of the thickness of
the $Ga_{1-x}Al_xAs$ layer.   The recombination velocity of 10^6 cm/sec
at the surface of the device would ordinarily result in a decreasing response at increasing photon energies, as shown in
Fig. 10, but the low recombination velocity at the interface
in the $Ga_{1-x}Al_xAs$ device eliminates the usual surface loss.
Light below 2.4 eV passes through the semitransparent $Ga_{1-x}Al_xAs$
layer and is absorbed in the underlying GaAs junction, which
has a recombination velocity at its "surface" (the interface)
of 10^4 cm/sec or less.   Above 2.4 eV, the response begins to
cut off due to absorption of light in the $Ga_{1-x}Al_xAs$.   As the
$Ga_{1-x}Al_xAs$ is made thinner, however, it absorbs less and the
cut off is moved to higher energies.   If the thickness is

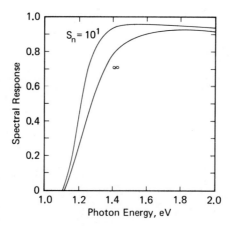

FIG. 15. *Computed internal spectral responses of N/P Si cells (1 ohm-cm, 4 mil) with either a BSF (S_{back} = 10) or an Ohmic back contact (S_{back} = ∞). Parameters of Table 4. No drift field in base.*

reduced still further, so that it becomes comparable to the electron diffusion length, some of the carriers generated in the $Ga_{1-x}Al_xAs$ will diffuse to the interface and add to the collection from the pGaAs, further increasing the response at high energies.

The low energy response from the base of Si cells can be improved by the addition of a back surface field (Fig. 7) as well as by a drift field close to the junction. The back surface field (BSF) configuration with a back region W_{p^+} much thinner than the main region W_p is equivalent in first approximation to a normal device with a very low recombination velocity at the back surface [43]. The effective low recombination velocity reduces the loss of photogenerated carriers that would ordinarily occur at the back contact, enhancing the spectral response at low photon energies as seen in Fig. 15 for a 4-mil thick Si cell. The improvement in the low energy spectral response by addition of the p^+ region is greater for thin cells than for thick ones; the influence of the back contact becomes small if the thickness of the base is much more than a diffusion length.

2. MEASURED RESPONSES

The calculated spectral responses of solar cells are in reasonable agreement with measured ones, even though many of the assumptions used in order to obtain analytical expressions

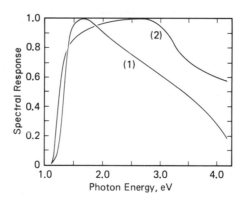

FIG. 16.   *Relative spectral responses (measured) of N/P Si solar cells.   (1) Low lifetime (dead layer), high surface recombination velocity device, with x_j = 0.3-0.4 µm [46]. (2) No dead layer, x_j = 0.1-0.2 µm, "violet cell" [4].*

are not accurate in describing real devices. Certainly the qualitative behavioral trends of measured devices having different lifetimes, diffusion lengths, electric fields, etc., are predicted very well by the analytical expressions, and comparing measured spectral responses of devices with ones predicted by theory is a very good method for studying solar cells and the effects of changes in design, fabrication processes, material parameters, and the like, on the device behavior. The wavelength dependence of the antireflection coating can easily be taken into account, or even ignored to a first approximation.

Spectral responses are measured using a monochromatic source of light, a detector, and a recording instrument. The light source can be a well-calibrated grating or prism spectrometer or a set of narrow bandpass filters, together with a high color temperature bulb. The detector is used to measure either the power of the monochromatic beam at each wavelength or to measure the number of photons in the beam directly; "black" detectors which need no correction for wavelength are the most convenient but any detector with a known wavelength response can be used. A recording must be made of both the detector output and test cell output at each wavelength to obtain the response of the cell. Low frequency beam-chopping and loc-in amplification are often used to obtain better signal-to-noise ratios at low intensities.

It has proven convenient in solar cell studies to normalize the measured spectral response to unity at the wavelength of maximum response; the result is called the relative spectral

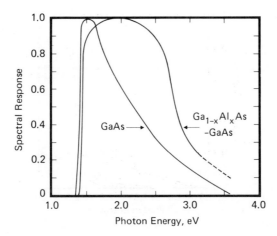

*FIG. 17. Relative spectral responses (measured) of a dif-
fused P/N GaAs solar cell and of a $Ga_{1-x}Al_xAs$-GaAs solar cell
($Ga_{1-x}Al_xAs$ thickness = 0.7 μm, x = 0.85).*

response. Then, if the absolute response (quantum efficiency)
is measured at any one wavelength, as with a laser, the absolute response at all wavelengths is obtained. When response
curves are published, it is usually the relative ones that are
given.

Figure 16 shows the relative spectral response as a function of photon energy of a standard type of 10 ohm-cm, diffused
Si N/P cell. The response begins at the bandgap energy, reaches
a peak at around 1.5 eV, and decreases with increasing energy
due to a combination of very short lifetime (perhaps less than
1 nsec) in the diffused region and a high surface recombination
velocity. An electric field may exist over at least a portion
of the diffused n-region, but it is apparently insufficient
to overcome the high losses. The spectral response of a "violet cell" as developed by Lindmayer and Allison [4] at Comsat
Laboratories is also shown in Fig. 16. In this cell the heavily damaged, nanosecond lifetime "dead layer" in the diffused
region has been largely eliminated by reducing the concentration of the phosphorus and by making the junction depth much
smaller than usual. The combination of higher lifetime near
the surface and narrower junction greatly improves the response
at high energies. Evidence of the saturation predicted by the
theoretical results (Fig. 8) can be seen.

Experimental results for GaAs and $Ga_{1-x}Al_xAs$-covered
GaAs P/N junctions are shown in Fig. 17. The GaAs device
shows the characteristic triangular-shaped response due to

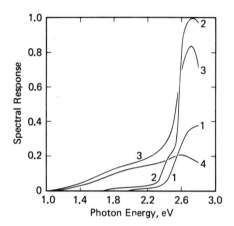

FIG. 18. *Relative spectral responses (measured) of Cu_2S-CdS thin film solar cells for increasing Cu_2S thickness (the numbers refer to increasing thickness). Light incident on Cu_2S surface (after Mytton [47], with permission from The Institute of Physics and the Physical Society, London).*

a high surface recombination velocity and a large junction depth to diffusion length ratio. The $Ga_{1-x}Al_xAs$ layer eliminates the surface losses but attenuates some of the high energy light due to absorption in the layer.

The spectral responses for front illumination of thin film polycrystalline CdS solar cells as presented by Mytton [47] are shown in Fig. 18. These devices are made by evaporating a thin CdS film onto a conducting substrate, plating a layer of copper onto the CdS surface, converting a thin region of the CdS to Cu_xS, and providing Ohmic contacts in grid form to the Cu_xS. The Cu_xS layer (usually 1000 to 3000 Å thick) has a bandgap of about 1.1 eV, and provides most or all of the response below the bandgap of the CdS (2.4 eV), while the CdS layer contributes most of the response above 2.4 eV.

The four curves in Fig. 18 show the effect of increasing the Cu_xS thickness by varying the plating times. If the Cu_xS is very thin, its transparency is too high and no low energy response is present (curves 1 and 2). If it is too thick (compared to a diffusion length), then carriers generated close to the Cu_xS surface may be lost, which decreases the response below 2.4 eV, while at the same time less response is obtained from the CdS above 2.4 eV because less light penetrates through to it (curve 4). The optimum Cu_xS thickness is around 2000 Å (curve 3), which results in some response from both the Cu_xS and CdS.

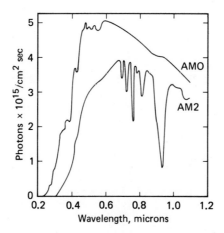

FIG. 19. Solar irradiance in photons per cm² per second in a 100 Å bandwidth for outer space (AM0) conditions and for average weather conditions on earth (AM2).

Higher responses between 1.1 and 2.4 eV are obtained when the Cu_xS layers are of higher quality in terms of diffusion length. Higher responses are also obtained in this range of wavelengths for "back-wall" cells where the device is constructed in such a way that it can be illuminated through the CdS, but the response above 2.4 eV is reduced in these cells because of attenuation by the CdS itself.

D. Short Circuit Current

1. CALCULATED PHOTOCURRENT

In addition to its value as a tool in studying solar cells, the spectral response can be used to compute the expected short circuit photocurrent for any given spectral input. Since the spectral response represents the number of carriers collected per incident photon, the photocurrent density per unit bandwidth at a given wavelength is given by

$$J_{ph}(\lambda) = qF(\lambda)SR(\lambda) \tag{35}$$

where the external response, Eq. (33), which includes the reflection of light from the surface, is used. The total photocurrent density obtained when sunlight (or any other light source) with a spectral distribution $F(\lambda)$ is incident on the cell is found by integrating (35)

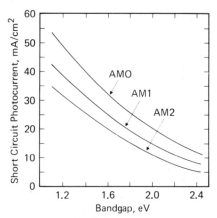

FIG. 20.   Idealized short circuit current densities as a function of semiconductor bandgap for unity spectral response.

$$J_{ph} = q \int_0^\infty F(\lambda) SR(\lambda) \, d\lambda \qquad (36)$$

This relationship is applicable as long as the excess minority carrier density generated by the light is small compared to the majority carrier density in the device, so that the differential equations involved in the spectral response remain linear [48]. (Low injection level conditions apply up to at least 10-20 solar intensities in 10 ohm-cm Si cells and 100 in GaAs devices; the higher the base doping level is, the greater will be the input intensity for which low injection level calculations are still valid.)

The solar irradiance in terms of the number of photons contained in a 100 Å bandwidth located at a wavelength λ, for wavelengths from 0.2 to 1.2 µm, is shown in Fig. 19. The higher curve represents sunlight outside the earth's atmosphere, the lower one represents the light received at the earth's surface on an average, nearly cloudless day. The degree to which the atmosphere affects the sunlight received at the surface is defined quantitatively by the "air mass." Technically, the air mass is equal to the secant of the angle of the sun to the zenith, measured at sea level, or in other words the path length that a ray of sunlight must traverse compared to the shortest path it could take. This definition is inadequate to describe the real situation on any given day, since it does not take into account the prevailing weather conditions on that day, but it serves at least as a quantitative estimate of what might be expected on the average at a given point at a given time of the year.

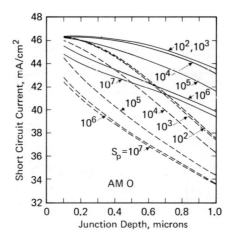

FIG. 21. Calculated AM0 short circuit photocurrents as a function of junction depth both with (solid) and without (dashed) an electric field in the top region. Poor conditions (dead layer) have been assumed for the top region. No field in the base. $S_{back} = \infty$. N/P Si cell: 1 ohm-cm, 18 mil.

The air mass 0 (AM0) spectrum in Fig. 19 represents the most accurate present estimate by NASA [49,50] for sunlight outside the earth's atmosphere, with a total incident power integrated over all wavelengths of 135.3 mW/cm^2 at the earth's distance from the sun. The AM2 spectrum [51] represents the sunlight at the earth's surface when the sun is at an angle of 60°, leading to a total incident power of 72-75 mW/cm^2; it also approximates the spectrum obtained for average, slightly hazy weather conditions and smaller sun angles. The AM1 spectrum represents the sunlight at the earth's surface for optimum weather conditions with the sun at the zenith, leading to a total incident power of slightly over 100 mW/cm^2, with a curve that lies between the curve for AM0 and that for AM2 in Fig. 19. The major differences between sunlight in space and the light received at the earth's surface are in the ultraviolet and infrared contents. Ultraviolet light is filtered out by ozone in the upper layers of the atmosphere, and infrared is removed from the spectrum by water vapor and CO_2. Aerosol particles scatter light of short wavelengths more than long wavelengths. The greater the number of atmospheric constituents there are, the more sunlight tends to be "channeled" in wavelength into the visible region, and the more the amplitude is attenuated at any wavelength. The diffuse component of light (light reaching the surface from all parts of the sky) is also increased relative to the direct component as the air mass increases.

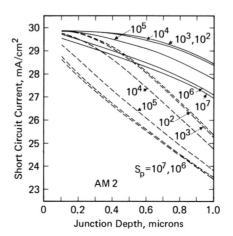

FIG. 22. Calculated AM2 short circuit photocurrents as a function of junction depth both with (solid) and without (dashed) an electric field in the top region. Same conditions as Fig. 21.

Materials with high bandgaps yield higher open circuit voltages than materials with lower bandgaps, but they also yield lower photocurrents because the sunlight at low energies (long wavelengths) is not absorbed. Figure 20 shows the highest current that could be obtained as a function of the bandgap, i.e., the idealized photocurrent that would be obtained if the absolute spectral response were equal to unity for all photon energies above the bandgap and zero for all energies below it. At AM0 the current decreases with bandgap from about 54 mA/cm^2 for Si to about 39 mA/cm^2 for GaAs and 14 mA/cm^2 for GaP. At AM2 the currents are 34, 25, and 6.5 mA/cm^2 for Si, GaAs, and GaP, respectively. The AM1 photocurrents are roughly halfway between the values calculated at AM0 and AM2.

The short circuit current is always less than the idealized values in Fig. 20 in a real situation because of losses due to bulk and surface recombination. Reducing the surface recombination velocities at the front and back, and improving the diffusion lengths in both the top region and the base would reduce these losses and bring measured photocurrents closer to the idealized values given above, but this is easier said than done; these parameters are largely determined by the properties of the material and by the procedures used to fabricate the device, and are usually not as good as one would like them to be.

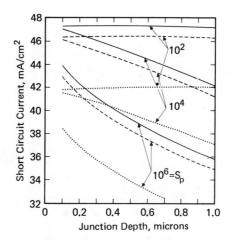

FIG. 23. Calculated AM0 short circuit photocurrents as a function of base resistivity and junction depth. Solid lines: 10 ohm-cm bases; dashed lines: 1 ohm-cm bases; dotted lines: 0.1 ohm-cm bases. No dead layer, no drift fields. Parameters of Table 4. $S_{back} = \infty$. *N/P Si cell: 18 mil.*

Two ways to minimize these recombination losses and improve the photocurrent are to decrease the junction depth and to provide electric drift fields, as has already been discussed for the spectral response. Figures 21 and 22 show the calculated photocurrent obtained at AM0 and AM2, respectively, for Si N/P cells with a 1 ohm-cm base resistivity using the material parameters of Table 4 (except for the hole lifetime and diffusion length in the top region, which were taken as 3 nsec and 0.62 μm to correspond to the low values (dead layer) commonly measured in phosphorus-diffused devices). The photocurrent is considerably improved by making the junction narrow, and further improved by the electric field (which is assumed to be equal to the difference in Fermi levels, E_V-E_F, at the surface and at the junction edge divided by the junction depth). High surface recombination velocities and low lifetimes in the top region become less important as the junction depth is decreased, partially because of reduced losses and partially because of the greater percentage of the photocurrent contributed by the base and depletion regions compared to the top region.

The photocurrents are larger for high base resistivities, where diffusion lengths and lifetimes are larger, than for low base resistivities where the diffusion lengths and lifetimes are reduced by the ionized impurity scattering. Figure 23

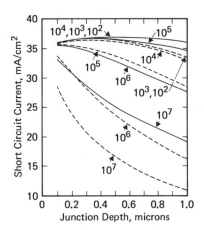

FIG. 24. GaAs P/N. Calculated AM0 short circuit photocurrents
as a function of junction depth and surface recombination velocity both with (solid) and without (dashed) an electric field
in the top region. Parameters of Table 5. Device thicknesses
of 15 μm or more. No field in the base. $S_{back} = \infty$.

shows the currents calculated for 10, 1, and 0.1 ohm-cm N/P
Si cells as a function of junction depth using the parameters
of Table 4 but without any drift fields. Photocurrent calculations for P/N devices fall below those of Fig. 23 by several
milliamperes per cm^2 because of the smaller hole diffusion
length in the base of P/N cells for an equivalent doping level
compared to the electron diffusion length in the base of an
N/P device.

     Note that the predicted photocurrent is independent of
junction depth when the dead layer is absent and when the surface recombination velocity is low. Under these conditions,
the junction depth could be made large (5 μm) as in epitaxial
growth of the top region without seriously affecting the photocurrent.

     The photocurrents obtained at AM0 and AM2 for GaAs P/N
solar cells both with and without electric fields in the top
region, using the parameters listed in Table 5, are presented
in Figs. 24 and 25. The same curves are obtained for overall
device thicknesses from 15 μm to infinity, and for back surface
recombination velocities from zero to infinity, since all the
carriers are created within the first 3 to 4 μm from the surface and since hole diffusion lengths are at best around 2.5-
3 μm for common base doping levels. Reducing the junction
depth and providing an electric field in the top region are
even more important for GaAs solar cells than for Si cells

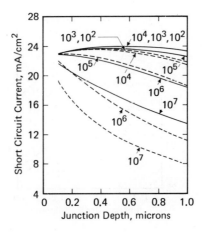

FIG. 25.   GaAs P/N.   Calculated AM2 short circuit photocurrents as a function of junction depth both with (solid) and without (dashed) an electric field in the top region.   Same conditions as in Fig. 24.

because of the larger percentage of carriers generated close to the surface in GaAs where the high recombination velocity and low lifetime usually result in severe losses.

The photocurrents obtained at AM0 and AM2 for $Ga_{1-x}Al_xAs$-GaAs solar cells as a function of junction depth and $Ga_{1-x}Al_xAs$ thickness are shown in Figs. 26 and 27. The underlying GaAs p-n junction is assumed to have the same parameters as in Table 5; the $Ga_{1-x}Al_xAs$ is assumed to have a recombination velocity of 10^6 cm/sec at its surface, and the recombination velocity at the interface is taken as 10^4 cm/sec (same conditions, except for junction depth, as in Fig. 14). The effect of the junction depth is less than for normal GaAs cells because surface recombination losses are effectively absent. The slight increase in photocurrent that does occur with decreasing junction depths (but above 0.4 μm) is a result of reduced bulk losses in the pGaAs region; the decrease in photocurrent for junction depths below 0.4 μm is caused by the loss of a few carriers generated at long wavelengths at distances of 3 to 4 μm from the surface. The strongest effect on the photocurrent is obtained when the $Ga_{1-x}Al_xAs$ thickness is decreased, since more light reaches the GaAs p-n junction under these conditions (and since some of the carriers generated in the $Ga_{1-x}Al_xAs$ can be collected). The benefit of having the $Ga_{1-x}Al_xAs$ layer on the GaAs surface compared to the results obtained without this layer can be seen by comparing Figs. 26

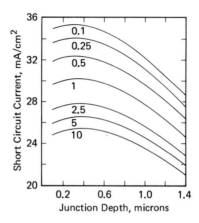

FIG. 26. $Ga_{1-x}Al_xAs$-GaAs. Calculated AM0 short circuit photocurrents as a function of junction depth and $Ga_{1-x}Al_xAs$ thickness. No drift fields. Same conditions as Fig. 14.

and 27 with 24 and 25; the photocurrents obtained from $Ga_{1-x}Al_xAs$ devices for thicknesses of 2.5 μm or less are higher than those obtained from GaAs devices with the same surface recombination velocity (10^6 cm/sec).

2.   EXPERIMENTAL PHOTOCURRENT

Almost all of the data available on short circuit currents are for AM0 sunlight, due to the much higher interest in space applications compared to terrestrial uses up to a few years ago. The standard 10 ohm-cm N/P Si cell found on most satellites yields [52] about 35 mA/cm^2, which becomes 38 mA/cm^2 after correction for the portion of the cell masked by the contacts and 41 mA/cm^2 after correcting for reflection, in good agreement with Fig. 21. N/P devices with 1 ohm-cm bases yield around 35-37 mA/cm^2 after both corrections (30-32 mA/cm^2 before any corrections). P/N Si cells with the same resistivities yield comparable but just slightly smaller currents; the smaller minority carrier diffusion length in the base of a P/N cell compared to an N/P device with the same base resistivity (see Table 4) is partly compensated for by the larger diffusion length expected in the top region of the P/N device. The "violet cell," with its improved spectral response at short wavelengths compared to conventional cells, yields [4, 53] around 40 mA/cm^2 uncorrected, or 46.5 mA/cm^2 after both

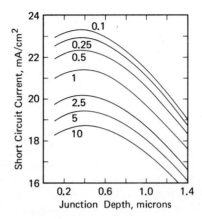

FIG. 27. $Ga_{1-x}Al_xAs$-GaAs. Calculated AM2 short circuit photo-currents as a function of junction depth and $Ga_{1-x}Al_xAs$ thickness. Same conditions as Fig. 26.

corrections, which is higher than in conventional 10 ohm-cm N/P devices even though the base in the violet cell is usually more heavily doped (2 ohm-cm) [53].

Very few measurements have been reported for GaAs p-n junctions. Gobat and co-workers [6] measured devices with fairly deep junctions (1 μm or over) in 1962 and obtained a short circuit current of 17.5 mA/cm^2 for AM1 after correcting for contact area (but not for reflection); this should translate to about 20 mA/cm^2 at AM0. Improvements in GaAs technology and the incorporation of drift fields in the top region to aid photogenerated carrier collection should readily increase this to 25 mA/cm^2 at AM0.

$Ga_{1-x}Al_xAs$-GaAs devices have been measured both in simulated and actual sunlight, and photocurrents (corrected for contact area) of 22.5 mA/cm^2 at AM0 [9], 21.3 mA/cm^2 at AM1 [8], and 19.5 mA/cm^2 [8] at AM2 were reported. The photocurrents in these devices were limited by attenuation in the $Ga_{1-x}Al_xAs$ layer, and higher currents will very likely be reported as the layer is made thinner and its Al content is increased.

The uncorrected short circuit currents of laboratory Cu$_2$S-CdS solar cells lie in the range of 25-30 mA/cm^2 at AM0 [54] and around 20-25 mA/cm^2 at AM1. The measured spectral responses of these devices were not given but must have been considerably better than those of Fig. 18 to yield such high currents. Production line devices have photocurrents around 2/3 to 3/4 of these values, and spectral responses more comparable to Fig. 18.   The photocurrents are due mostly to collection from the Cu_2S with its 1.1 eV bandgap, and to a lesser degree, to collection from the CdS with its 2.4 eV bandgap.

E.   Summary

Two of the most important parameters in a solar cell are the minority carrier lifetime and the minority carrier diffusion length.   In the base of the cell, the lifetime and diffusion length depend on the method of growing the crystal, the procedures used to prepare the substrate, the substrate resistivity, the presence of undesirable impurities such as oxygen and copper, and the annealing temperatures and times (if any). In the very thin top region, which is usually prepared by diffusion, the lifetime and diffusion length depend on the type of dopant and its surface concentration and the surface treatment prior to the diffusion.   High surface concentrations and their associated stress, dislocations, and perturbations of the band structure can lead to a "dead" layer of extremely low lifetime over a fraction of the diffused top region adjacent to the surface.

The ability of a solar cell to generate photocurrent at a given wavelength of incident light is measured quantitatively by its "spectral response," and the total short circuit photocurrent obtained from the cell is the product of the spectral response and the number of photons in the incident light spectrum, integrated over all wavelengths.   Solar cell spectral responses and photocurrents have been computed for three device models, including uniformly doped base and top regions, constant electric fields in both regions, and a back surface field (blocking back contact).

The response at long wavelengths depends mostly on the base lifetime and diffusion length.   If the diffusion length is low, the response can be improved by the presence of an electric field in the base adjacent to the depletion region, but if the diffusion length is high a base field will have small effect.   A blocking back contact can improve the response by minimizing the number of photocarriers that would ordinarily recombine at an Ohmic back contact.   The back contact conditions are only important if the device thickness is less than several base diffusion lengths.

The response at short wavelengths depends mostly on the front surface recombination velocity and the lifetime in the top region.   "Dead" layers and high recombination velocities greatly lower the response, while reducing the junction depth and providing an electric field in the top region are beneficial in overcoming these problems. Eliminating the dead layer entirely by optimizing the diffusion conditions and lowering the recombination velocity by surface passivation or other techniques are important in obtaining high short circuit currents.

All solar cells are affected by the conditions in both the top region and the base. Solar cells made from indirect gap materials such as Si, however, are more dependent upon the conditions in the base, while devices made from direct gap materials such as GaAs are governed more by the conditions of the top region. Maximum theoretical short circuit currents of 54 and 39 mA/cm^2 at AMO and 34 and 25 mA/cm^2 at AM2 are predicted for Si and GaAs solar cells, respectively, but actual photocurrents are usually about 2/3 of these values due to the recombination losses and reflection of light and contact area losses. The measured photocurrents of Cu_2S-CdS cells at AMO have reached 25 mA/cm^2; most of this current is due to collection from the Cu_2S layer.

CHAPTER 3

Solar Cell Electrical Characteristics

A. Current Mechanisms

The voltage-current behavior of a solar cell in the dark is equally as important as the photocurrent in determining the output of the cell, since the junction characteristics determine how much of the electrical energy developed by the cell with light incident on it will be available at the output terminals and how much will be lost as heat. When power is being taken from the cell, a voltage exists across its terminals in the forward bias polarity, and a junction "dark current" exists which is opposite in direction to the photocurrent. The current being supplied to the load is the photocurrent minus this dark current, and it is important to have as low a dark current as possible at the operating voltage to obtain the highest efficiency.

In all p-n junctions, several current transport mechanisms (transport of holes and electrons across the depletion region) can be present at the same time, and the magnitude of each one is determined by the doping levels on the two sides of the junction and by the presence of any added energy barriers as in heterojunctions. Such transport mechanisms in the forward bias direction include injection of carriers over the junction barrier, recombination of holes and electrons within the depletion region, and injection of carriers up a portion of the barrier followed by tunneling into energy states within the bandgap (tunneling may take place through a series of steps with recombination in between, as in the "excess" current in tunnel diodes). These currents are shown diagramatically in Fig. 28; there may be other possibilities in special cases.

In the absence of shunt or series resistance effects, the dark I-V characteristics of a solar cell are given by the sum of the current mechanisms that are present. In a normal Si p-n junction device with 1 or 10 ohm-cm base material, the tunneling current is not likely to be of importance compared

48

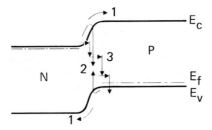

FIG. 28. *Three current transport mechanisms in forward biased p-n junctions: (1) injection; (2) recombination within the depletion region; and (3) multistep tunneling via energy states caused by defects.*

to the other two, but in Si devices made with 0.01 ohm-cm bases, the tunneling component will probably dominate. For 0.1 ohm-cm devices (a few times 10^{17} cm^{-3}) the tunneling current should be negligible for devices with high junction perfection but both the tunneling and depletion region recombination mechanisms will be increasingly important as the number of defects (and hence energy states) within the depletion region increases. These defects can be caused by impurities, dislocations resulting from stress caused by the junction diffusion, and other problems introduced during the device fabrication. (Heterojunctions and Schottky barrier devices tend to be subject to tunneling currents more than p-n homojunctions are. Tunneling has been suggested by Fahrenbruch and Bube [55] and by Böer and Phillips [56] as one of the two major dark current components in Cu_2S-CdS cells.)

In this section, each of these three dark current components will be described in turn. The injected current will be described for each of three device models: uniformly doped regions, constant electric fields, and a back surface field. The space charge layer recombination current and the tunneling component are largely independent of the model assumed; they depend mostly on the doping levels at the edges of space charge region.

1. INJECTED CURRENTS

The dark current-voltage relationships in p-n junctions are derived from equations very similar to (6) and (7). The injected current component (1 in Fig. 28) consists of electrons injected from the n-side over the potential barrier into the p-side, where they diffuse and drift (if there is an electric field) away from the junction and eventually recombine either in the bulk or at a surface. The current component also consists of an analogous current due to holes injected from the p-side into the n-side. The behavior of these minority carriers is governed by the continuity equations

$$(1/q)(d/dx)J_n-[(n_p-n_{p0})/\tau_n] = 0 \qquad \text{(electrons on p-side)},$$

$$\qquad\qquad\qquad\qquad\qquad\qquad\qquad\qquad\qquad\qquad\qquad (37)$$

$$(1/q)(d/dx)J_p+[(p_n-p_{n0})/\tau_p] = 0 \qquad \text{(holes on n-side)},$$

$$\qquad\qquad\qquad\qquad\qquad\qquad\qquad\qquad\qquad\qquad\qquad (38)$$

and by the current equations

$$J_n = q\mu n_p E+qD_n(d/dx)n_p \qquad \text{(p-side)}, \qquad\qquad\qquad (39)$$

$$J_p = q\mu p_n E-qD_p(d/dx)p_n \qquad \text{(n-side)}. \qquad\qquad\qquad (40)$$

In order to obtain an analytical result, it is necessary to assume all the parameters (μ,τ,E,D) to be constant; if these assumptions cannot be used even as a first approximation, numerical methods can be used [39], which are more accurate but far less expedient.

The boundary conditions necessary for the solution of (37)-(40) in an N/P cell are

$$p_n = p_{n0}\exp(qV_j/kT) \qquad (x = x_j), \qquad\qquad\qquad (41)$$

$$n_p = n_{p0}\exp(qV_j/kT) \qquad (x = x_j+W), \qquad\qquad\qquad (42)$$

$$S_p(p_n-p_{n0}) = D_p(d/dx)(p_n-p_{n0})-\mu_p p_n E_p \qquad (x = 0), \qquad (43)$$

$$S_n(n_p-n_{p0}) = -D_n(d/dx)(n_p-n_{p0})-\mu_n n_p E_n \qquad (x = H), \qquad (44)$$

where $x = 0$ is the front of the cell and $x = H$ is the back. Equations (41) and (42) are the Boltzmann relationships for carriers on the two sides of the junction when the voltage across the junction is V_j, i.e., V_j is the voltage introduced across the junction either by light or by some other means such as a battery.

The maximum value that the photovoltage can theoretically have is the "built-in" potential V_d, which is related to the bandgap by

$$V_d = E_g-(E_C-E_F)-(E_F-E_V) \qquad\qquad\qquad\qquad (45)$$

$$\quad = (kT/q)\,\ln(N_aN_d/n_i^2). \qquad\qquad\qquad\qquad\qquad (46)$$

The closer the Fermi levels lie to their respective band edges on the two sides, the higher the potential V_d will be; for degenerate conditions V_d can even exceed the bandgap. (The actual open circuit voltage V_{oc} is always less than V_d; V_{oc} is equal to the voltage at which the photocurrent is exactly opposed by the total dark current, and if the dark current components are large at a given voltage, then V_{oc} will be correspondingly small.)

The injected current component, which is by far the largest in normal Si solar cells, can now be found by solving (37) through (44) under various conditions. The solutions will be obtained for N/P cells, but analogous expressions apply for P/N devices.

a. Uniform Doping

If the doping levels on the two sides are constant, then the electric fields outside the depletion region are negligible and the solution takes a simple form. One form of the solution to (37)-(40) can be written as

$$p_n - p_{n0} = A_1 \cosh(x/L_p) + B_1 \sinh(x/L_p) \qquad (x \leq x_j) \qquad (47)$$

$$n_p - n_{p0} = A_2 \cosh\left[(x-(x_j+W))/L_n\right]$$
$$+A_2 \sinh\left[(x-(x_j+W))/L_n\right] \quad (x_j+W \leq x \leq H) \qquad (48)$$

and using the boundary conditions (41)-(44), the injected current becomes

$$J_{inj} = J_0 (\exp[qV_j/kT] - 1) \qquad (49)$$

where

$$J_0 = q \frac{D_p}{L_p} \frac{n_i^2}{N_d} \left[\frac{(S_p L_p/D_p) \cosh(x_j/L_p) + \sinh(x_j/L_p)}{(S_p L_p/D_p) \sinh(x_j/L_p) + \cosh(x_j/L_p)}\right]$$
$$+q \frac{D_n}{L_n} \frac{n_i^2}{N_a} \left[\frac{(S_n L_n/D_n) \cosh(H'/L_n) + \sinh(H'/L_n)}{(S_n L_n/D_n) \sinh(H'/L_n) + \cosh(H'/L_n)}\right] \qquad (50)$$

where $H' = H-(x_j+W)$ and S_p, S_n are the recombination velocities at the front and back of the cell, respectively.

b.   Uniform Electric Fields

When doping gradients exist, electric fields will be present outside the depletion region. If these fields are assumed to be constant across the device, then the preexponential J_0 in Eq. (49) for an N/P cell becomes

$$J_0 = q \, \frac{D_p}{L_{pp}} \, \frac{n_i^2}{N_d} \left[\frac{\sinh(x_j/L_{pp}) + ((S_p L_{pp}/D_p) + E_{pp}L_{pp}) \, \cosh(x_j/L_{pp})}{((S_p L_{pp}/D_p) + E_{pp}L_{pp}) \, \sinh(x_j/L_{pp}) + \cosh(x_j/L_{pp})} \right.$$

$$\left. - E_{pp}L_{pp} \right] + q \, \frac{n_i^2}{N_a} \left[\frac{D_n}{L_{nn}} \, \coth \frac{H - (x_j + W)}{L_{nn}} - E_{nn}D_n \right] \tag{51}$$

as presented by Ellis and Moss [37]. The terms E_{pp}, L_{pp}, E_{nn}, and L_{nn} were defined in Eqs. (26) and (27). The first term represents current injected from the diffused n-region into the base and the second term represents current injected from the base in the opposite direction. An infinite surface recombination velocity (as for a metallic Ohmic contact) was assumed at the back of the cell [37].

c.   Back Surface Field

In Godlewski's et al. [43] treatment of the current flow when a back surface field (p$^+$-region) is present (Fig. 7), the three neutral regions of the device are uniform in doping level, lifetime, and mobility (no drift fields) and the dark current component from the base is then given by [43]

$$J_0 \text{(base)} = q \, \frac{D_n}{L_n} \, \frac{n_i^2}{N_a} \left[\frac{(SL_n/D_n) \, \cosh(W_p/L_n) + \sinh(W_p/L_n)}{\cosh(W_p/L_n) + (SL_n/D_n) \, \sinh(W_p/L_n)} \right] \tag{52}$$

where

$$S = \frac{N_a}{N_a^+} \, \frac{D_n^+}{L_n^+} \left[\frac{(S_n L_n^+/D_n^+) \, \cosh(W_p^+/L_n^+) + \sinh(W_p^+/L_n^+)}{\cosh(W_p^+/L_n^+) + (S_n L_n^+/D_n^+) \, \sinh(W_p^+/L_n^+)} \right] \tag{53}$$

and N_a, D_n, L_n are the properties of the p-base region, while N_a^+, D_n^+, L_n^+ are those of the p$^+$-region. W_p and W_p^+ are the widths of the lightly and heavily doped base regions, respectively (Fig. 7). The base current (52) is of the same form as the second term in (50), except that the recombination

velocity "seen" by the electrons in the p-region is S instead
of S_n, and S can be much less than S_n, particularly if the
doping level N_a^+ is much greater than N_a (the effect of the
BSF is largest for high p-region resistivities and for narrow
widths W_p relative to the diffusion length L_n). This tendency
of the BSF to "confine" the minority carriers in the p-region
and away from the back contact can lower the dark current
through the device considerably and consequently improve the
open circuit voltage and fill factor.

The contribution to the dark current from the n^+-diffused
front region of a BSF cell is the same as given by the first
term of (50) or the first term of (51).

2. SPACE CHARGE LAYER RECOMBINATION CURRENT

When a p-n junction is forward biased, electrons from the
n-side and holes from the p-side are injected across the junction depletion region into the p- and n-sides, respectively,
but at the same time some of these carriers recombine inside
the depletion region, resulting in an increase in the dark
current through the device. This "space charge layer recombination current" was first discussed by Sah *et al.* [57] in
1957, and was later extended by Choo [58]. In the Sah-Noyce-
Shockley (S-N-S) theory, the doping levels were assumed to be
the same on the two sides of the junction and a single recombination center located in the vicinity of the center of the
gap was assumed also. The dark current component under forward
bias was derived as

$$J_{rg} = \frac{qn_iW}{\sqrt{\tau_{p0}\tau_{n0}}} \frac{2\sinh(qV_j/2kT)}{q(V_d-V_j)/kT} f(b), \tag{54}$$

where V_d is the built-in voltage, W is the depletion region
thickness, and τ_{p0}, τ_{n0} are the minority carrier lifetimes on
the two sides of the junction. The factor f(b) is a complicated expression involving the trap level E_t and the two lifetimes

$$f(b) = \int_0^\infty \frac{dx}{x^2+2bx+1} \tag{55}$$

$b = [\exp(-qV_j/2kT)] \cosh[(E_t-E_i)/kT+(1/2)\ln(\tau_{p0}/\tau_{n0})]$,

where E_i is the intrinsic Fermi level. The function f(b) has a maximum value of $\pi/2$, which occurs at small values of b (forward biases > 2kT/q); f(b) decreases as b increases.

Choo later extended the S-N-S theory to the more general case where the doping levels are not the same on the two sides, where the level E_t can be away from the gap center, and where the two lifetimes τ_{n0}, τ_{p0} can be orders-of-magnitude apart. He derived an equation virtually identical to (54) except that the function f(b) is smaller than in the S-N-S case, i.e., the effect of junction assymetries is to lower the recombination current below the value predicted by the S-N-S derivation. Equation (54) with f(b) = $\pi/2$ is therefore the largest value that the recombination current is expected to take, provided that the theories adequately describe the real situation [59].

3. TUNNELING CURRENT

A third type of dark current component that can exist under some situations is a tunneling current caused by electrons or holes tunneling from the conduction or valence band into energy states within the bandgap, followed by either tunneling the remainder of the way into the opposite band or by a tunneling-recombination mechanism (the current marked 3 in Fig. 28). Tunneling is not likely to be important in 10 and 1 ohm-cm Si cells, but in 0.01 ohm-cm Si devices, heterojunctions such as Cu_2S-CdS, and Schottky barriers, tunneling can be a major contributor to the dark current.

Heterojunctions offer a particularly graphic method for studying tunneling currents because they are very often dominated by tunneling, a result of the many energy states that can be introduced within the bandgaps by the lattice and thermal expansion mismatches and by the cross-doping of one material into the other. These tunneling currents take the form [60]

$$J_{tun} = K_1 N_t \exp(BV_j) \tag{57}$$

where K_1 is a constant containing the effective mass, built-in barrier, doping level, dielectric constant, and Planck's constant, N_t is the density of energy states available for an electron or hole to tunnel into, and B is a constant containing the doping level, dielectric constant, and the effective mass, $B = (4/3\hbar)(m^*\varepsilon/N_{d,a})^{1/2}$. This tunneling dark current (57) varies exponentially with voltage just as J_{inj} and J_{rg} do, and can easily be mistaken for one of these. The value of B (the slope of $\ell n\ J$ versus V) often lies in the range of 20 to 30 [61]; if this slope was mistakenly assumed to be qV_j/AkT, values for A of 1.3-2 would be deduced at room temperature. The only method for differentiating tunneling currents from thermal ones such as J_{inj} and J_{rg} is by temperature measurements; tunneling currents are very insensitive to temperature, while the opposite is true for thermal currents.

4.   TOTAL CURRENT

When more than one dark current component is present, the I-V characteristics are given by the sum of them

$$J_{dark} = J_{inj} + J_{rg} + J_{tun}. \tag{58}$$

In most Si, GaAs, and $Ga_{1-x}Al_xAs$-GaAs solar cells, only the first two will be important, but for Cu_2S-CdS, other types of heterojunctions, Schottky barriers, and very heavily doped p-n junctions, the tunneling current may also be important.

The major differences between the space charge layer recombination current and the injected current lie in their voltage, temperature, and bandgap dependences. For 1 to 10 ohm-cm Si devices, the value of J_{inj} extrapolated to zero bias is around 10^{-12} A/cm^2, while the value of J_{rg} is around 10^{-8} A/cm^2. At the same time, J_{inj} varies as $\exp(qV_j/kT)$ while J_{rg} varies as $\exp(qV_j/2kT)$, so that the recombination current dominates at low forward biases and the injected current dominates at higher biases, with a crossover at around 10^{-5} to 10^{-4} A/cm^2. J_{inj} has a bandgap dependence of $\exp(-E_g/kT)$, while J_{rg} varies as $\exp(-E_g/2kT)$; therefore, J_{rg} becomes increasingly important relative to J_{inj} for high bandgap materials and at low temperatures.

The injection and recombination currents for a Si solar cell with a 1 ohm-cm base resistivity and for two values of lifetime in the diffused n-region are shown in Fig. 29. The doping level is so much higher in the diffused region than in the base that the injection current is determined by electrons injected into the base alone to all practical purposes, and is independent of the top surface recombination velocity and the lifetime in the top region. This top region lifetime does have a significant effect on the recombination current, however, because of its appearance in the square-root radical of Eq. (54). If τ_{p0} is high, such as the saturation value of 0.4 µsec suggested by Ross and Madigan [18], J_{rg} is much smaller than J_{inj} at the operating voltage (0.4-0.5 V) and has little effect on the device behavior. If τ_{p0} is very low however, as is often measured in conventional cells after the phosphorus diffusion, then the recombination current is considerably larger; J_{rg} becomes comparable to or even larger than J_{inj} at the

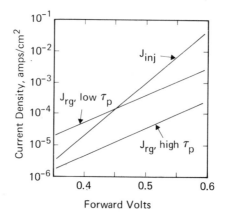

FIG. 29.   *Calculated injection and recombination currents in a Si N/P solar cell for the two cases of high and low lifetimes in the top region. Parameters of Table 4. No drift fields. S_{back} = ∞. Conditions: 1 ohm-cm, 450 µm thick, x_j = 0.3 µm.*

maximum power point (Fig. 2) and reduces both the open circuit voltage and fill factor.   (Whenever the defect density is high in the depletion region, and hence the lifetime is low there, J_{rg} can be expected to be unusually high.   This could particularly be a problem in ribbon Si devices and polycrystalline thin film cells, as suggested by Stirn [62].)

The recombination current in GaAs is much larger relative to the injection current than it is in Si, as can be seen by comparing Fig. 30 with Fig. 29.   The injection current in GaAs cells is still determined mostly by the base region due to the much lower doping level there.

The high value of the recombination current in GaAs devices is largely a result of the very low lifetimes in the two regions. J_{rg} is often much lower in LPE-grown devices, where the carrier lifetimes can be around 10^{-8} sec on both sides of the junction, than in bulk or vapor grown material where the lifetimes are usually smaller.

B.   Equivalent Circuit

The simplest equivalent circuit of a solar cell in the operating mode is shown in Fig. 31.   The photocurrent is represented by a current generator I_{ph} and is opposite in direction to the forward bias current of the diode $I_{inj}+I_{rg}$.   Shunt resistance paths are represented by R_{sh}; they can be caused by surface leakage along the edges of the cell, by diffusion

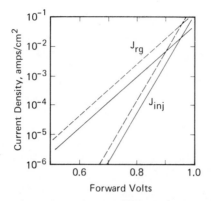

FIG. 30. *Calculated injection and recombination currents in*
GaAs P/N cells for both low losses $(S_{front} = 10^4$ *cm/sec,* $\tau_n =$
1×10^{-9} *sec,* $\tau_p = 1.6 \times 10^{-8}$ *sec; solid lines) and high losses*
$(S_{front} = 10^6$ *cm/sec,* $\tau_n = 1 \times 10^{-9}$ *sec,* $\tau_p = 2 \times 10^{-9}$ *sec; dashed*
lines). No drift fields. $S_{back} = \infty$. *Conditions: 0.01 ohm-cm;*
$H = 20$ *μm-∞;* $x_j = 0.5$ *μm.*

spikes along dislocations or grain boundaries, or possibly by
fine metallic bridges along microcracks, grain boundaries, or
crystal defects such as stacking faults after the contact
metallization has been applied. Series resistance, represented
by R_s, can arise from contact resistances to the front and back
(particularly for high resistivity bases, 1 to 10 ohm-cm), the
resistance of the base region itself, and the sheet resistance
of the thin diffused or grown surface layer. More complicated
equivalent circuits can be formulated to account more accu-
rately for the distributed nature of both the series resistance
and the current generator [63].

The dark currents I_{inj} and I_{rg} are equal to the current
densities J_{inj} and J_{rg} multiplied by the *total* device area A_t.
The photocurrent is equal to the photocurrent density multiplied by the *active* device area A_a, i.e., the total area minus
the area masked by the front contacts. The equivalent circuit
and the resulting relationships must be written in terms of
currents, not current densities. The difference between the
total area and the active area should be kept in mind, although
the difference is usually only 6-8% and is often neglected to
a first approximation.

From the equivalent circuit of Fig. 31, a relation can be
written between current output I_{out} and voltage output V_{out}.
Assuming the dark current to be $I_{inj} + I_{rg}$ as given by (49) and
(54) multiplied by the total area, this relation is

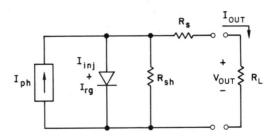

FIG. 31.   *Equivalent circuit of a solar cell, including series and shunt resistances.*

$$I_{out}(1+(R_s/R_{sh})) = I_{ph}-(V_{out}/R_{sh})-(I_{dark}) \tag{59}$$

where I_{inj} and I_{rg} are functions of the junction voltage V_j, which is in turn equal to $(V_{out}+I_{out}R_s)$.

1.   RELATIONSHIPS FOR NEGLIGIBLE R_s, R_{sh} LOSSES

In order to use the equivalent circuit to predict solar cell output and efficiency and to make the relationships analytically manageable, the approximations are often made that series and shunt resistance effects are negligible and that the dark current can be written as

$$I_{dark} = I_{00}[\exp(qV_j/A_0kT)-1], \tag{60}$$

where the "junction perfection factor" A_0 and the new value of the preexponential factor I_{00} have been used to approximate the sum of I_{inj} and I_{rg} by a single term (a glance at Fig. 29 will show that such an approximation is justified if the minority carrier lifetimes are high but not if they are low).   The advantage of these approximations is that (59) takes a very simple form

$$I_{out} = I_{ph}-I_{00}[\exp(qV_{out}/A_0kT)-1]. \tag{61}$$

A plot of (61) has already been shown in Fig. 2 for $A_0 = 1$; the effect of higher values of A_0 is to round out the "knee" in the curve near the maximum power point.

The short circuit current is given simply by

$$I_{sc} = I_{ph} \tag{62}$$

and the open circuit voltage by (1)

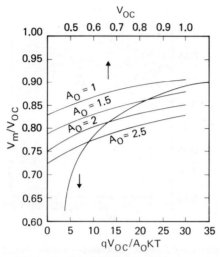

FIG. 32. *The ratio of the maximum power point voltage to the open circuit voltage as a function of the normalized open circuit voltage (unmarked curve), valid for all V_{OC}, A_0, and T. Also shown are four curves for $A_0 = 1$-2.5 and 298°K; the abscissa for these curves is at the top.*

$$V_{oc} = A_0(kT/q) \ln((I_{sc}/I_{00})+1)$$

$$= A_0(kT/q) \ln((J_{sc}A_a/J_{00}A_t)+1). \tag{1}$$

(It might be thought from (1) that high values of A_0 would be desirable in obtaining high open circuit voltages, but this is actually not the case, since high A_0 also requires high I_{00} to approximate the two currents I_{inj} and I_{rg} by the one term (60); V_{oc} for p-n junctions is always higher for low values of A_0 than the opposite [64].)

Since the power output is $V_{out} \times I_{out}$, the maximum power output can be found by differentiating the product and setting the result equal to zero [64,65]

$$P_{out}(max) = I_m V_m \tag{63}$$

where

$$I_m = (I_{sc}+I_{00}) \left[\frac{qV_m/A_0 kT}{1+(qV_m/A_0 kT)} \right] \tag{64a}$$

$$= (J_{sc}A_a+J_{00}A_t) \left[\frac{qV_m/A_0 kT}{1+(qV_m/A_0 kT)} \right] \tag{64b}$$

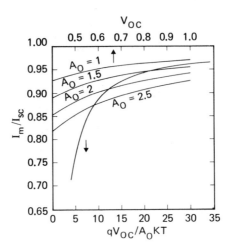

FIG. 33.   The ratio of the maximum power point current to the short circuit current as a function of the normalized open circuit voltage, valid for all V_{oc}, A_0, and T.   Also shown are four curves for $A_0 = 1$-2.5 at $298°K$.

is the current output at maximum power and

$$\exp(qV_m/A_0kT) \cdot [1+qV_m/A_0kT]$$

$$= (I_{sc}/I_{00})+1 = \exp(qV_{oc}/A_0kT) \tag{65a}$$

$$= ((J_{sc}A_a/J_{00}A_t)+1) \tag{65b}$$

allows the voltage at maximum power output to be calculated. The fill factor (FF), which is $V_mI_m/I_{sc}V_{oc}$, measures the "squareness" of the I-V curve, and is found to be [66]

$$FF = V_m\left[\frac{1-(I_{00}/I_{sc})[\exp(qV_m/A_0kT)-1]}{(A_0kT/q)\ \ell n[I_{sc}/I_{00})+1]}\right] \tag{66a}$$

$$= V_m\left[\frac{1-(J_{00}A_t/J_{sc}A_a)[\exp(qV_m/A_0kT)-1]}{(A_0kT/q)\ \ell n[(J_{sc}A_a/J_{00}A_t)+1]}\right] \tag{66b}$$

$$= \frac{V_m}{V_{oc}}\left[1-\frac{[\exp(qV_m/A_0kT)-1]}{[\exp(qV_{oc}/A_0kT)-1]}\right] \tag{66c}$$

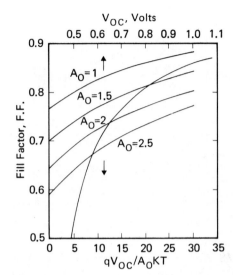

FIG. 34. Fill factor as
a function of normalized
open circuit voltage, valid
for all V_{OC}, A_0, and T.
Also shown are four curves
for A_0 = 1-2.5 at 298°K.

The relationships contained in (64) to (66) are shown in
Figs. 32-34, assuming that the total and active areas are equal.
(It should be kept in mind that these relationships are only
useful in the idealized case where there are no series or shunt
resistance effects and where the current can be represented by
the single exponential of Eq. (60).)

The two ratios V_m/V_{OC} and I_m/I_{SC} and the FF all improve
with increasing values of V_{OC} and with decreasing values of A_0
and T. Higher bandgap materials yield higher ratios and fill
factors because of their higher open circuit voltages (provided
series and shunt resistances are not a problem).

The nearer the value of A_0 is to unity, the better the
device performance is, other things being equal. In Si, for
example, with a V_{OC} of around 0.58 V, FF is equal to 0.82 for
A_0 = 1 but only 0.72 if A_0 = 2. For GaAs with V_{OC} = 0.9 V, FF
decreases from a potential value of 0.87 for A_0 = 1 to 0.79
if A_0 = 2.

2. EFFECTS OF R_{sh} AND R_s

When series and shunt resistance problems become important,
the relationships of (64)-(66) no longer apply; the two ratios
V_m/V_{OC} and I_m/I_{SC} and the FF are all reduced below the values
shown in Figs. 32-34. The relationship in (59) between V_{out}
and I_{out} becomes almost impossible to solve analytically, although a numerical solution can be readily obtained. The
effects of series and shunt resistances on solar cell behavior

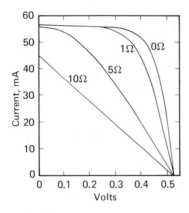

FIG. 35.   The effects of series
resistance on measured Si solar
cell curves.   Tungsten light,
$100\ mW/cm^2$.   Cell area $= 2\ cm^2$.

can be seen easily by placing various resistors alternately
in series and in parallel with an otherwise normal solar cell.
Figure 35 shows the effect of series resistance on the output
of a commercial Si cell illuminated with tungsten light at
$100\ mW/cm^2$ intensity.   The open circuit voltage is not changed
but the fill factor is seriously reduced.   There can also be
a reduction in the short circuit current below the value of the
photocurrent due to the forward bias across the diode caused
by the voltage drop across the series resistance (even though
the total output voltage is zero) which results in appreciable
dark current in opposition to the photocurrent.   Even small
values of series resistance, in the 0.5 to 1.0 ohm range for
$2\ cm^2$ cells, are enough to cause serious effects.

Figure 36 shows the effect of shunt resistances in parallel
with the solar cell (same device as in Fig. 35); in this case
the short circuit current is not affected, but the fill factor
and open circuit voltage are reduced as the shunt resistance
decreases.

In practical devices, the shunt resistance is usually
large enough to have a negligible effect at 1 solar intensity
or above.   At low intensities however, and to some degree at
low temperatures, the shunt resistance takes on increasing
importance [23].   On the other hand, the series resistance
becomes increasingly important at high intensities and temperatures.   The need to minimize the series resistance suggests
high doping levels and deep junctions which are just the opposite of the necessary conditions for high current collection
efficiency.   The compromise has been reached to make the diffused region thin but very highly doped, and at the same time,
to optimize the design of the Ohmic contact grid pattern [67,
68] for the lowest sheet resistance consistent with covering
only 5-10% of the surface.   With the commonly found six-finger

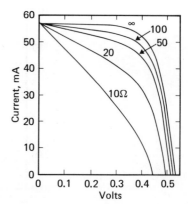

FIG. 36. The effects of shunt resistance on measured Si solar cell curves. Same conditions as Fig. 35.

grid pattern used so much in the past, the series resistance can be as much as 0.5 ohm for a 2 cm^2 Si cell; at the operating current of 60 to 65 mA, 30 to 33 mV can be lost across this resistance. Increasing the number of fingers while decreasing the finger width and the distance between fingers lowers the series resistance and the voltage loss [4]. This has become particularly important for devices such as the "violet cell" that have 1000-2000 Å junction depths and lower doping levels in the diffused region. Violet cells are designed with 30 fingers/cm [4], with a final contact area equal to 6-7% of the total. The resulting series resistance is around 0.05 ohm for a 4 cm^2 device, considerably less than the 0.2-0.25 ohm of more conventional Si devices, in spite of the higher sheet resistivity and narrower width of the diffused region.

C.   Experimental Current-Voltage Behavior

Measured Si devices almost always show the effects of both series and shunt resistances and have higher recombination currents than predicted by theory [23,69]. Figure 37 shows the dark I-V measurement of a 10 ohm-cm N/P Si solar cell. The current at low voltages (less than 0.1 V) is due to shunt resistance (around 10^5 ohms). Two exponential regions can be seen starting at 0.2 V, with slopes of qV/2.1kT and qV/1.1kT, respectively. The decreasing slope of the current around 3 mA/cm^2 is a result of the series resistance of the device, around 1 ohm. At the short circuit current value of 30 mA/cm^2, 30 mV are lost across this resistance.

Nearly all large-area Si cells show some shunt resistance effects, a fact which is difficult to explain from ordinary theories. Edge leakage is one source of low shunt resistances.

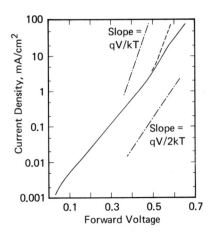

FIG. 37.   Dark I-V characteristics of a commercial Si solar cell (at 300°K), showing the two exponential regions and the effect of series resistance. (The dashed line is the characteristic after correcting for the series resistance.)

It is difficult to etch the edges of a large-area Si device and finish off the etching by a technique that results in a low density of surface states at the device edges; contamination from the chemicals used and water vapor included in the oxide that forms on the edges can both result in leakage.   With proper passivation, though, edge leakage can be minimized.

Stirn has pointed out [23] that shunt resistance problems can arise from small scratches and imperfections on the device surface which become partially or totally covered by the contact metallurgy during the device fabrication; "sintering" the contact stripes to minimize contact resistance can cause small metal particles to enter the scratch and result in leakage across the p-n junction.   Since the scratches (and possibly other imperfections such as stacking faults) are random along the surface, certain areas of the device should be better in electrical properties than others.   Stirn [23] has demonstrated this by comparing the I-V characteristics of a 2×2 cm commercial cell with the characteristics of small mesas etched on the same device (Fig. 38); most small mesas show almost negligible shunting compared to the full device, while some mesas exhibit much higher leakage than the average.   The leakage current of the full cell is sometimes increased by up to a hundredfold after contact sintering compared to before sintering, further establishing the role of the metallization in causing shunt resistance problems.

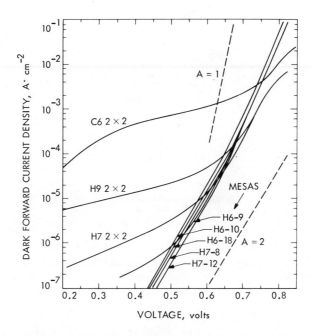

FIG. 38. Dark I-V characteristics of Si solar cells (2 ohm-cm N/P, 192°K), showing the high leakage current measured in complete (4 cm^2) cells compared to mesas etched on the same cells. The A refers to the slope, from qV/AkT. (After Stirn [23]; courtesy of the IEEE.)

Most Si and GaAs solar cells exhibit several exponential regions in the dark forward I-V characteristics, as seen in Fig. 37 for a Si device and Fig. 39 for a GaAs device; this strongly suggests the presence of several current components such as J_{inj} and J_{rg}. Very seldom do the slopes of these exponentials (the value of A in qV/AkT) equal 1 or 2. For good Si devices, values close to unity are observed (1.1-1.3) at high voltages and close to 2 (1.6-1.8) at lower voltages, with the smallest values corresponding to the best devices. For poor devices A values of 3 or even 4 are observed. Values of the saturation current J_{00} (Eqs. (60)(61)) are around 10^{-10} to 10^{-9} A/cm^2 when the high current (low A) part of the I-V curve is estrapolated to zero volts, and 10^{-7} to 10^{-6} A/cm^2 for the A $\sim$ 2 portion. According to theory, these should be around 10^{-12} to 10^{-11} A/cm^2 for J_{inj} and 10^{-8} A/cm^2 for J_{rg}, for 10 ohm-cm cells.

Values of A close to 1 are almost certainly due to a dominance of the injected current J_{inj} (of all the possible current components, this seems to be the only one capable of

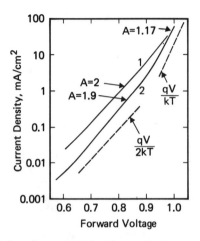

FIG. 39. Dark I-V characteristics of GaAs solar cells (P/N, 300°K). (1) Device with low lifetimes and diffusion lengths; (2) device with high lifetimes and diffusion lengths. Both curves have been corrected for series resistance.

yielding a value of 1). At high injection levels, amperes per cm^2 rather than milliamperes per cm^2, A approaches 2, even for the injected component, but this current level is not reached in solar cells unless they are operated at several hundred solar intensities. Values of A close to 2 at low injection levels are most likely due to space charge region recombination J_{rg} or to recombination at the edges of the device within the space charge region [70]; both of these mechanisms can exhibit A's in the range of 1 to 2 under certain conditions [70]. Sah [70] has measured the I-V characteristics of a number of diffused Si junctions with small areas and found three well-defined regions when series resistance is absent; at low voltages A values of 1.2-1.4 are observed and attributed to the presence of both J_{inj} and J_{rg}. At medium voltages (0.3-0.5 V in Sah's devices) the injected current becomes dominant and A falls to around 1. At high voltages (0.7-0.8 V in Sah's devices) high injection levels are reached and A rises to 2. The higher the doping level is in the base, the higher the voltage "threshold" is at which high injection level effects begin to take place.

High values of A (>2) are not predicted by the S-N-S theory [57], but could be due in part to shunt resistance effects [23] (since a low value of shunt resistance causes a shallow slope in the I-V curve which can be mistaken for a recombination current with a high value of A) and in part to modifications in the S-N-S theory which account for nonuniformities in the distribution of recombination centers [70,71]. Shockley and

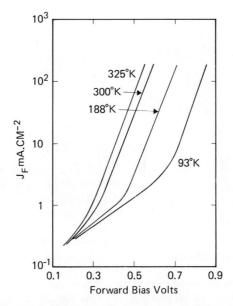

FIG. 40.   Dark I-V characteristics of a thin film Cu_2S-CdS solar cell as a function of temperature.   The constant slope and small change in magnitude with temperature imply tunneling. (After Martinuzzi et al. [77]; courtesy of the IEEE.)

co-workers [71,72] attributed high values of A in their Si junctions to a reduced density of recombination centers at the middle of the space charge region compared to points away from the middle. Nakamura and co-workers [73] have found that re-distribution of heavy metal impurities (gettering) can take place during device processing with strong increases of impurity densities near the surface; this would tend to increase measured recombination currents and measured values of A above values predicted by theory. Sah has postulated that high A values can arise in planar p-n junction devices due to surface channels caused by surface states. In solar cells these channels would lie along the device edges and extend into the base.

   GaAs solar cells tend to be dominated by recombination currents and most devices exhibit a qV/2kT dependence over much of their current range. The dominance of the recombination component is most likely due to low values of lifetime and dif-fusion lengths; if lifetimes corresponding to 4 μm diffusion lengths are used in Eqs. (50) and (54), theory predicts that J_{rg} will dominate at low voltage (<0.8 V) and J_{inj} will dominate at higher values. Figure 39 shows the dark I-V curves of two GaAs p-n junction solar cells (pGa$_{1-x}$Al$_x$As-pGaAs-nGaAs devices)

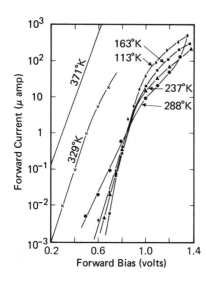

FIG. 41.   Dark I-V character-istics of a single-crystal Cu_2S-CdS solar cell after normal heat treatment, 1 min at 250°C.   (After Gill and Bube [74]; courtesy of the American Institute of Physics.

produced by liquid-phase epitaxy.   Device #1 is known to have diffusion lengths of around 1 μm and exhibits a qV/2kT varia-tion over the entire range of measured currents.   Device #2 is estimated from spectral response measurements to have diffu-sion lengths of 3 μm or more; it exhibits recombination current with A = 1.9 at voltages less than 0.9 V and injection current with A = 1.17 at higher voltages.   Device #1 is typical of vapor diffused GaAs cells and device #2 is typical of good LPE-produced units.

The dark current-voltage characteristics of heterojunction devices are nearly always dominated by tunneling.   Many experi-menters have observed tunneling currents in Cu_2S-CdS solar cells in both the forward and reverse-biased directions [55,74-78]. Figures 40 and 41 show dark forward I-V characteristics of such cells as a function of temperature; the independence of the slope of the ℓn J versus V and the small change in the current magnitude with temperature are characteristic of tunneling-limited currents.   The I-V characteristics are affected strongly by the heat treatment normally employed during the device pro-cessing; the current after heat treatment measured at room temperature and below is decreased by several orders-of-magni-tude below its preheated value.   Gill and Bube [74] have sug-gested that deep acceptor imperfections (probably Cu ions) have diffused for a short distance into the normally n-type CdS during the heat treatment, widening the depletion layer and lowering the tunneling probability.   In later papers, Fahren-bruch, Lindquist, and Bube [55,75,76,78] suggest that the dark current after heat treatment consists of both a thermal injec-tion component and a tunneling component.   The injection current becomes dominant above 320°K and is due to the injection of electrons from the CdS into the Cu_2S conduction band; the activation energy of 1.2 eV for this current corresponds to the barrier height between the two conduction bands. The tunneling current is dominant below 320°K (as shown by Fig. 41) and is assumed to be caused by the tunneling of electrons from the CdS into interface states where they recombine with holes which have tunneled there from the Cu_2S. The observed activation energy of 0.45 eV for the tunneling is due to the need for the electrons to thermally surmount a portion of the energy barrier before the remainder of the depletion region in the CdS becomes thin enough for appreciable tunneling to occur.

The current in this tunneling regime can be described by [55,76,78]

$$J = J_0 \exp(BV) \tag{67}$$

where the J_0 term is highly dependent on processing because of the variability of defect states (N_t in Eq. (57)) while the exponent B is dependent on the doping levels, dielectric constants, and number of intermediate tunneling steps. The constant B is not strongly affected by processing as is J_0 [78]; typical values for B for non-heat-treated devices (for voltages above 0.35 V) range from 24 to 30, while J_0 varies from 10^{-9} A/cm^2 to almost 10^{-6} A/cm^2 [76]. For heat-treated units (several minutes to several hours at 100-250°C), B is again about 24 to 30, while J_0 has decreased by 2 to as much as 5 orders-of-magnitude [74]. The thermal component with 1.2 eV activation energy which is present in heat treated devices above 320°K is not observed in non-heat-treated units [74] due to the much higher tunneling current in these units that essentially "swamps out" any injection current that might be present.

D. Summary

The current-voltage behavior of a solar cell in the dark is just as important as its behavior in the light, since the dark behavior largely determines the voltage output and fill factor. The dark I-V characteristics are determined by the combined effects of the current transport mechanisms which may be present and any series and shunt resistance problems that may arise. The two current components of most importance are the injected component due to the injection of minority carriers from the top region into the base, and the depletion region recombination current due to the recombination of partially injected holes and electrons within the depletion region.

In cases of high doping levels ($>10^{18}$ cm^{-3}) in both the base and top regions, a third component due to tunneling may be present.

The injected current is determined mostly by conditions in the base, and has been calculated for the three models of uniform base doping, constant electric field in the base, and a back surface field. The depletion region recombination current and the tunneling current are determined by conditions within the depletion region, and depend strongly on the width of this region, the lifetime within it, and the number of defect states available for tunneling.

Experimental I-V measurements indicate that the depletion region recombination current is considerably higher in Si and GaAs solar cells than expected from theory. This might be attributed to a poor lifetime in the depletion region due to unwanted impurities introduced during the diffusion, or it might indicate that the present theory of depletion region recombination requires revision to bring it closer to the experimental results. Shunt resistance problems can add to the difficulty of interpreting I-V data, but these problems can be minimized with proper care during contact sintering and with care in preventing scratches and other defects from being introduced during processing.

The equivalent circuit of a solar cell consists of a photocurrent generator in parallel with a diode and a shunt resistance, and a series resistance leading to the output terminals. From this equivalent circuit, the power output from the cell can be calculated under various conditions. The series resistance lowers the short circuit current without affecting the open circuit voltage, while the shunt resistance does just the opposite. Both resistances degrade the fill factor.

Analytical expressions can be derived for the fill factor and for the voltage and current operating points if the series and shunt resistance effects can be ignored and if the I-V characteristic can be represented by a single exponential instead of the sum of several exponentials. Fill factors of 0.75 to 0.82 for Si cells and 0.79 to 0.85 for GaAs cells are predicted this way. Usually series resistance (and sometimes shunt resistance) effects cannot be ignored, and several current mechanisms are present. This leads to slightly lower fill factors and operating voltages, in agreement with experimental results.

CHAPTER 4

Efficiency

The efficiency of a solar cell in converting sunlight
into useful electrical energy is the single most important
number defining the quality of the cell. Unfortunately, there
has been no clear-cut standardization of solar cell efficiency
measurements over the years, and different numbers have been
reported for the same type of cell without clearly stating the
spectral conditions during the measurement. A problem arises
due to the nonuniform spectral responses of solar cells; they
convert light of some wavelengths better than they do other
wavelengths. Since the solar spectrum outside the earth's
atmosphere is different from the spectrum received on the earth
on a clear day, and both of these differ from the spectrum on
a hazy day, the efficiencies measured under each of these conditions are different. In the early days of solar cells the
spectral conditions for outdoor measurements were often not
even mentioned, and there was undoubtedly some error in many
of the reported values. In some instances efficiencies were
reported for tungsten bulb (indoor) incident light, which has
a relatively low color temperature compared to sunlight. Today,
very good simulation of outer space (AM0) sunlight has been
developed, consisting of a xenon light source with certain
types of absorbing filters to remove certain peaks in the xenon
line spectrum. A good simulator for earth sunlight at AM1 or
AM2 has yet to be developed, but a reasonable simulation can
be obtained with a quartz-halogen bulb and a 2-3 mm water filter.
Another source of confusion in solar cell measurements
arises from the tendency of some experimenters to report values
corrected for the contact area loss and others to report uncorrected values. Uncorrected efficiencies are reported on
the philosophy that these describe what the cell can actually
deliver, while efficiency values corrected for the contact area
are reported on the basis that these are the inherent efficiencies of the devices without the human factors of contact design
and process technology. Actually, it seems reasonable that

both values should be reported at the same time, which would
eliminate ambiguity.

High photocurrents, open circuit voltages, and fill factors naturally lead to high efficiencies in solar cells. A
wide, flat spectral response in the visible and near-infrared
spectral regions and a peak quantum efficiency close to unity
lead to high photocurrents, while low forward dark currents
and high shunt resistances lead to high open circuit voltages.
Good fill factors can be obtained if the forward dark currents
are low, the value of A_0 (the diode I-V "perfection factor")
is low, the series resistance is low (less than 1 ohm for a
1 cm^2 area), and the shunt resistance is high (greater than
10^4 ohms). Materials with high bandgaps theoretically have
higher open circuit voltages and fill factors while lower bandgap materials yield higher photocurrents, leading to a maximum
in the efficiency versus bandgap at about 1.5 eV.

In this chapter, the efficiencies at AM0, AM1, and AM2
will be described for several device models and for various
device parameters, as has been done for the short circuit current and dark current in the previous two chapters. Dead
layers, high surface recombination velocities, poor lifetimes,
and high series or shunt resistance losses result in low efficiencies. Reducing the junction depth, incorporating aiding
electric fields, improving the lifetimes, and preventing resistance losses are all beneficial in improving the efficiencies.

A. Calculated Efficiencies

The efficiency of a solar cell in converting light of any
arbitrary spectral distribution into useful power is given by

$$\eta = V_m I_m / P_{in} \tag{68}$$

where V_m, I_m are the voltage and current at the maximum power
point (Fig. 2). The input power is

$$P_{in} = A_t \int_0^\infty F(\lambda)(hc/\lambda)\,d\lambda \tag{69}$$

where A_t is the total device area, $F(\lambda)$ is the number of photons
per cm^2 per sec per unit bandwidth incident on the device at
wavelength λ and hc/λ is the energy carried by each photon.
For sunlight, the spectrum $F(\lambda)$ was shown in Fig. 19 at AM0 and
AM2. The power output is given by

$$P_{out} = V_m I_m \equiv FF\ V_{oc} I_{sc}. \tag{70}$$

These equations have been written in terms of current rather than current density to take account of the small difference between the total and active device areas A_t and A_a.

It is possible to find analytical expressions for the efficiency under certain idealized conditions, namely, those for which the series and shunt resistance losses are ignored. In this case, assuming the dark forward I-V characteristic can be approximated by the single exponential in Eq. (60), the efficiency can be written as

$$\eta = \frac{FF(A_0kT/q)\ \ell n((I_{sc}/I_{00})+1)qA_a\int_0^\infty F(\lambda)SR(\lambda)_{ext}\ d\lambda}{A_t\int_0^\infty F(\lambda)(hc/\lambda)\ d\lambda} \tag{71}$$

where FF is given by (66), V_{oc} by (1), and I_{sc} by (36) multiplied by the active area.

Historically, a second type of analytical expression has been derived [64,65] under the same idealized assumptions which places the efficiency in terms of the average number of carriers collected and the average energy of the photon in the spectrum

$$\eta = Q(I_m/I_{sc})(V_mqn_{ph}(E_g)/N_{ph}E_{av})(A_a/A_t)(1-R) \tag{72}$$

where N_{ph} is the total number of photons per cm^2 per sec in the source spectrum, E_{av} is their average energy, $n_{ph}(E_g)$ is the number of photons per cm^2 per sec with energy greater than the bandgap, R is the average reflectivity, and Q is the average "collection efficiency," the ratio of the number of carriers collected to $n_{ph}(E_g)$, the number capable of being collected. The collection efficiency is related to the spectral response by

$$q Q n_{ph}(E_g)(1-R) = q\int_0^\infty F(\lambda)SR(\lambda)_{ext}\ d\lambda = J_{sc} \tag{73}$$

and in the event that monochromatic light is incident, the spectral response and the collection efficiency are identical. (The two ways of expressing the efficiency are, of course, equivalent, but the second method involves a number of averages and is more difficult to work with.)

The efficiency is a function of the bandgap through the influences of V_{oc}, FF, and I_{sc}. It was conjectured in the early solar cell days that a maximum would exist at some bandgap between 1.0 and 2.0 eV. In order to establish what this optimum bandgap might be, a quantity known as the "limit

conversion efficiency" was calculated by making the ideal
assumptions of 100% absorption of all photons with energies
greater than the bandgap, 100% collection of all generated
carriers, and *ideal* junction characteristics (along with the
assumptions of negligible series and shunt resistance effects,
negligible contact area, and negligible light reflection).
Under these conditions, the efficiency can be written as

$$\eta (ideal) = \frac{FFA_a qn_{ph}(E_g)(kT/q) \; \ell n\left[(qn_{ph}(E_g)/J_0)+1\right]}{A_t \int_0^\infty F(\lambda)(hc/\lambda)\; d\lambda} \quad (74)$$

where A_a and A_t are equal. The ideal short circuit current
density, $qn_{ph}(E_g)$, has already been shown in Fig. 20 and the
limit conversion efficiencies at AM0 and AM2 calculated from
(74) are shown in Fig. 42. The maximum occurs at around 1.5 eV
at 22.5% for AM0 and around 1.4 eV at 26% for AM2. The shape
of the AM2 curve and the higher efficiencies at AM2 are due
to the removal of most of the ultraviolet and portions of the
infrared light by the atmosphere, channeling the sun's energy
more and more toward the visible region where the spectral
response is high as the air mass increases.

It is important to regard calculations such as those of
Fig. 42 in a qualitative sense only. In order to obtain this
curve, which demonstrates the effect of bandgap alone, it is
necessary to assume that material parameters such as mobilities,
lifetimes, doping levels, and densities of states are the same
for all materials over the applicable bandgap range. For Fig.
42 the parameters of 1 ohm-cm Si have been used, but if the
values applicable to 0.1 ohm-cm Si were used and 100% collection still assumed, both of the curves in Fig. 42 would be
shifted upward. Different diffusion lengths and lifetimes,
densities of states, doping levels, recombination velocities,
and the directness or indirectness of the energy bandgap, all
have strong effects on the expected efficiencies of real devices
and individual real materials such as GaAs, GaP, InP, etc.,
may fall either above or below such an idealized curve.

The most accurate way to calculate the actual efficiencies
expected under various conditions is to use numerical methods
as outlined by Fossom [39]. The continuity and current equations together with Poisson's equation are solved exactly,
taking into account the variation of lifetime, mobility, electric field, and majority and minority carrier densities as a
function of doping level and position. Fossom's type of analysis is particularly valuable when used to calculate high injection level behavior, where the generated minority carrier

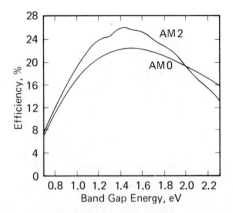

FIG. 42. *Limit conversion efficiencies as a function of energy gap at AM0 (input power density = 135.3 mW/cm^2) and at AM2 (input power density ≈ 74 mW/cm^2).*

densities become comparable to the majority carrier densities (several hundred solar intensities in GaAs, 50 to 100 in Si).

It is much easier, however, to use analytical tools than it is to use numerical methods. A reasonably accurate method of calculating expected efficiencies for a wide range of conditions consists of the use of (36) to compute the photocurrent and (58) to compute the dark current as a function of voltage. Equation (59) yields the relationship between I_{out} and V_{out} for arbitrary series and shunt resistances, and the efficiency is just the maximum in $V_{out} \cdot I_{out}$ divided by the solar input. Reflection of incident light from the surface as a function of wavelength and losses due to contact area can also be included as desired.

If series and shunt resistance effects can be ignored, the output current is given simply by $(I_{ph}-I_{dark})$, and the power output is just $V_j(I_{ph}-I_{dark})$. Since the series and shunt resistances and the contact area loss are determined largely by technology (series resistance and contact loss can be minimized by optimizing the grid design and shunt resistance can be maximized by edge passivation and by preventing metal-semiconductor interdiffusion during contact sintering), and since reflection of the incident light can be made small by proper antireflection coatings, it has proven useful to calculate the somewhat idealized efficiencies obtained by neglecting these technology-oriented losses. The result will be an "inherent" device efficiency demonstrating the effects of dead layers, depletion region recombination currents, surface recombination,

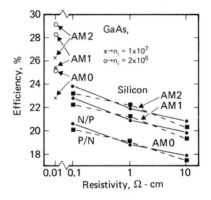

FIG. 43. *Calculated inherent efficiencies of Si N/P (solid) and P/N (dashed) solar cells versus base resistivity under optimistic conditions. The efficiencies for GaAs P/N cells under optimum conditions are also shown. $T = 300°K$, $S_{front} = S_{back} = 10^2$ cm/sec, drift fields present.*

junction depth, electric fields, and lifetime. It is assumed that light makes only one pass through the device, and exits at the back surface rather than being reflected back into the cell; such multiple passes of light would raise the predicted efficiencies slightly, and can become important for thin solar cells where the absorption of long wavelength light in a single pass can be low.

1. SILICON

In this section the "inherent" efficiencies of Si solar cells calculated by the method just outlined will be shown.

The efficiencies of Si N/P and P/N solar cells under the most optimistic conditions are shown in Fig. 43 for AM0, AM1, and AM2. These are the counterparts to the "limit conversion efficiencies," i.e., they are computed using the highest lifetimes found in the literature for Si, those of Kendall [79]. AM0 efficiencies of 18 to 21% could conceivably be obtained if these very high lifetimes of the best bulk Si could also be obtained in finished devices (τ = 200, 50, 20, 500, 200, and 50 μsecs for minority carriers in 10, 1, and 0.1 ohm-cm p-type and n-type Si, respectively).

The lifetimes measured in actual Si devices have generally been about an order-of-magnitude less than these optimistic values, and the expected efficiencies of actual devices are correspondingly less. Some of the expected material parameters for good Si devices were shown in Table 4, and the efficiencies

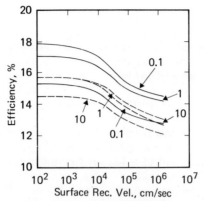

FIG. 44. Calculated inherent efficiencies at AM0 of N/P
(solid) and P/N (dashed) Si solar cells for x_j = 0.4 μm and
the parameters of Table 4. The numbers refer to the base
resistivity in ohm-cm. No drift fields, no dead layer.
S_{back} = ∞. Conditions: high τ_n, τ_p; 18 mil.

for AM0 of 10, 1, and 0.1 ohm-cm N/P and P/N devices calculated
using these parameters are shown in Fig. 44 as a function of
the front surface recombination velocity. The lifetime in the
top region has been taken at a high value as given by Ross and
Madigan [18] (τ_{p0} = 0.4 μsec, τ_{n0} = 1.1 μsec), i.e., there is
no "dead layer" present. For the best case under these conditions, that of a 0.1 ohm-cm N/P device, the highest calculated
efficiency is close to 18% (values for 0.01 ohm-cm devices have
not been computed because of the uncertainty in the effect of
the tunneling current J_{tun} at high doping levels in Si). The
fill factor and open circuit voltage improve while the photocurrent decreases with increasing doping level in the base;
the net result of these conflicting factors is that the efficiency improves somewhat with decreasing base resistivity.
The efficiencies of N/P devices are slightly higher than those
of P/N cells for equal resistivities, due to the higher doping
level for p-type material compared to n-type and due to the
larger lifetimes and diffusion lengths for electrons compared
to holes for equal resistivities.
     So far, the efficiencies of solar cells under rather good
conditions have been discussed. It is also important to dis-
cuss the difficulties that can arise and what can be done about
them. There are two serious loss mechanisms that may be present
in the top region: surface recombination, and bulk recombination in a "dead layer" of very low lifetime. The effect of
these two losses on the spectral response and dark current behavior of Si devices has already been discussed in Chapters 2

and 3, and the effect of one of them (surface recombination)
for cells with the parameters of Table 4 has been included in
Fig. 44. Surface recombination is caused by surface states
which result from unsatisfied bonds and from impurities present
at the surface; recombination velocities can be as high as 10^5-
10^6 cm/sec on Si, and 10^6-10^7 cm/sec on GaAs. A "dead layer"
in the top region is a region near the surface with nanosecond
or even subnanosecond lifetime, caused by the strain, dislocations, and unwanted impurities sometimes introduced during the
diffusion or during other processing steps. Such dead layers
have been described particularly for phosphorus-diffused devices [4], and the success of the "violet cell" has been attributed to the elimination of the dead layer [4] by lowering the
surface concentration of the diffusion and reducing the junction
depth.

These recombination losses in the top region can be reduced
by decreasing the junction depth and by incorporating an aiding
drift field in the top region; both of these steps improve the
photocurrent and the efficiency. Decreasing the junction depth
minimizes the number of carriers generated in the top region
and moves the edge of the depletion region (a perfect "sink"
for minority carriers) closer to the surface, so that generated
carriers have a higher probability of reaching it rather than
recombining. An aiding electric field improves the collection
efficiency by adding a drift force on the photogenerated carriers, moving them toward the junction.

The beneficial effects of a reduced junction depth and
an aiding drift field in the top region can be seen in Fig. 45.
These efficiencies have been calculated for a 1 ohm-cm N/P device with a dead layer present; the average lifetime in the top
region has been taken as 3 nsec. The combined effects of surface recombination and the dead layer can lower the efficiency
substantially for large junction depths, but lowering the depth
(which also increases the magnitude of the electric field)
brings the efficiency nearly to as high a value (17%, Fig. 44)
as can be obtained for low values of S_p and without a dead
layer present.

The highest predicted AM0 efficiencies are obtained for
0.1 ohm-cm p-type bases (see Fig. 44) when there is no dead
layer present and when the junction depth is low and the drift
field high to overcome surface recombination. Values of nearly
18% can be obtained for the parameters of Table 4 even with
relatively high values of S_p (but no dead layer), as shown in
Fig. 46 (solid lines), and over 17% can be obtained even with
a velocity of 10^6 cm/sec. The efficiencies could be even higher
if the base lifetime were larger, since insufficient collection
of photogenerated carriers in the base is a serious loss at

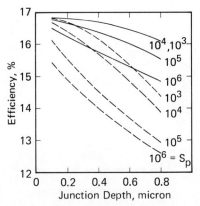

FIG. 45. *Calculated inherent efficiencies of a 1 ohm-cm Si N/P solar cell at AMO with a dead layer present in the top region ($\tau_p = 3\times10^{-9}$ sec). Solid lines—with drift field present in top region; dashed lines--without a drift field. H = 18 mil. $S_{back} = \infty$.*

these high doping levels. The computed efficiencies for 0.1 ohm-cm devices with the parameters of Table 4 except for the base lifetime and diffusion length are shown in Fig. 46 as dashed lines (lifetime = 10 μsec, diffusion length = 104 μm). Efficiencies of nearly 20% are predicted under the best conditions, and almost 19% even for a high value of S_p. (The presence of a dead layer in the top region would lower these calculated values by about 1%, i.e., from 20 to 19%.) The higher base lifetime improves the short circuit current, but it improves the open circuit voltage even more by reducing the dark current.

Theoretically, the improvement shown for the high base lifetime can also be obtained by incorporating a drift field in the base. Quantitatively, however, it is not clear how beneficial a base drift field might be on the efficiency, since the increasing doping level in the base as a function of position, necessary to obtain the drift field, also results in a decreasing lifetime and mobility as a function of position in the base. It seems likely that base drift fields could result in improved efficiencies, at least for low and moderate base lifetimes, but the difficulty and cost of diffusing over distances of tens of microns might outweigh the benefits obtained.

It should be noted that slightly higher efficiencies could be predicted for 0.01 ohm-cm resistivities than for higher base resistivities if the tunneling component is ignored, but the validity of ignoring this current is doubtful at the 10^{19} cm^{-3} doping levels present at this resistivity.

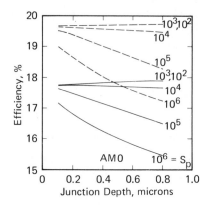

FIG. 46.   Calculated inherent efficiencies of a 0.1 ohm-cm Si N/P solar cell at AM0 without a dead layer present and with an electric field in the top region.   Solid lines—expected good base lifetime of 2.5 μsec; dashed lines--very high base lifetime of 10 μsec.   H = 18 mil. S_{back} = ∞.

Figure 46 also shows that the efficiency is constant as a function of junction depth when the surface recombination velocity is low and the bulk lifetime in the top region is high. The efficiency continues to be constant for junction depths up to about 7-8 μm, after which the bulk recombination becomes significant. It should be possible therefore to fabricate efficient Si solar cells by epitaxial growth of the top region rather than by diffusion, as long as the thickness of the epitaxial layer is no more than about 2/3 of the value of the diffusion length in the top region.

The effect of the "back surface field" as far as the efficiency is concerned for these 450-μm thick devices is practically (but not totally) negligible. The back surface field concept becomes particularly valuable when the thickness is reduced, as discussed in Chapter 5.

2.   GALLIUM ARSENIDE

The inherent AM0, AM1, and AM2 efficiencies of GaAs P/N cells under optimistic conditions are shown in Fig. 43.   There are two sets of numbers shown; the first (X's) represent values obtained if n_i, the intrinsic carrier density, is equal to $1×10^7$ cm^{-3} as is commonly assumed for GaAs [80].   The second set of values (circles) represent numbers computed if n_i = $1.8×10^6$ cm^{-3} as measured by Sell and Casey [81].   Efficiencies of 22-25% are predicted at AM0, 25-28% at AM1, and 26-29% at

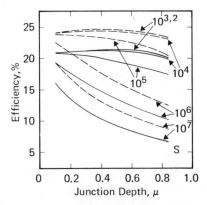

FIG. 47. Calculated inherent efficiencies of 0.01 ohm-cm GaAs solar cells at AM0 (solid) and AM2 (dashed) for the parameters of Table 5. No drift fields. $S_{back} = \infty$. H = 12 mil.

AM2 under idealized conditions. The lifetimes used in these calculations were 2.1×10^{-8} sec for the base and 4×10^{-9} sec for the top region, with S at the front surface equal to 10^3 cm/sec or less and an aiding drift field present (the recombination velocity at the back surface has no effect on the calculations at these device thicknesses).

The lifetimes measured in good finished devices made from high quality starting material are almost as high as those assumed in the optimistic calculation of Fig. 43 (see Table 3). The surface recombination velocity, on the other hand, is generally much worse than 10^3 cm/sec, often about 10^6-10^7 cm/sec for finished devices. This high surface loss is much more drastic for GaAs and other direct bandgap solar cells than it is for Si because light is absorbed and carriers created much closer to the surface. The high recombination velocity reduces the short circuit current by lowering the collection efficiency, and can even lower the open circuit voltage and fill factor by increasing the forward dark current (the portion of the forward dark current arising in the top becomes significant if the ratio of the doping level in the top region to the doping level in the base becomes less than 10).

The efficiencies calculated at AM0 and AM2 of GaAs P/N cells with various surface recombination velocities and junction depths (but no drift fields) are shown in Fig. 47 for lifetimes of 1×10^{-9} sec in the top region and 1.58×10^{-8} sec in the base (Table 5). A decrease in recombination velocity from 10^7 to 10^5 cm/sec increases the efficiency by a factor of 2 1/2 at large junction depths (7-17.5% at AM0) and a further decrease to 10^4 cm/sec would raise the efficiency by a

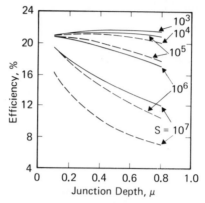

FIG. 48.   *Calculated inherent efficiencies of 0.01 ohm-cm
GaAs P/N solar cells at AM0, both with (solid) and without
(dashed) an electric field in the top region. Parameters of
Table 5.* $S_{back} = \infty$.   *H = 12 mil.*

factor of almost 3 (7-20%). Reducing the junction depth helps
considerably; the AM0 efficiency can reach 16% for $S = 10^7$ cm/sec
and 19% for $S = 10^6$ cm/sec at 0.1 μm junction depths even without
aiding fields. (The lifetime and mobility in the base become
increasingly important as the junction depth is reduced, however.
For example, if the diffusion length in the base is small, 1 μm
or less, long wavelength-generated carriers will be lost as the
junction depth is reduced and decreasing the depth will only
give small improvement in the overall photocurrent. If the base
diffusion length is large, 2 μm or more, decreasing the depth
will not have much effect on the long wavelength collection but
will strongly improve the response to short wavelengths, giving
a large improvement to the overall photocurrent.)

Since the surface losses are so drastic in GaAs, it stands
to reason that aiding drift fields in the top region could be
of considerable benefit in improving the device behavior, even
more than in Si devices. This was first pointed out by Ellis
and Moss [37] in 1970, who predicted 20-21% AM0 efficiencies
for N/P cells with narrow junctions. The same improvement can
be obtained for P/N devices with aiding drift fields, as shown
in Fig. 48. The electric field nearly doubles the efficiency
of cells with deep junctions and recombination velocities of
10^7 cm/sec, and cells with narrow junctions can reach AM0 efficiencies of nearly 20% even though the recombination velocity
is this high. The field helps considerably in improving the
efficiencies of GaAs P/N or N/P devices with other values of
surface recombination velocity also.

There are two other features of GaAs solar cells which should be mentioned before going on. The first is that the problem of a "dead layer" is probably not as important as in Si solar cells, where the lifetime in the top region is some-times several orders-of-magnitude less than it ought to be at the doping level which is present. In GaAs devices, the life-time is already so low to begin with that it takes a relatively great deal of lattice damage to lower it much further, and in any case, diffusions (and vapor growth) of GaAs are carried out at lower temperatures than for Si, and the lattice damage intro-duced should be correspondingly lower. Very good GaAs solar cells can be made with electron lifetimes of 0.5 to 1 nsec in the top region (as in Table 5 and Figs. 47 and 48), and fairly good devices can theoretically be made with electron lifetimes as low as several hundred picoseconds, provided the hole life-time in the *base* is 5 to 10 nsec.

The second feature is the change in the optimum design from n-type bases to p-type when the junction depth is reduced to less than 0.1 µm. This is due to the larger electron dif-fusion length compared to the hole diffusion length at a given doping level. When the junction depth is made very small, ≤0.1 µm, the base region becomes more important than the top region and it is better (i.e., higher efficiencies are obtained) if electrons are the minority carriers in the base. If the junction depth is larger than around 0.1 µm, though, the very small diffusion length for holes (0.1 µm) at the high doping levels found in the diffused top region of N/P devices will result in a high loss of short wavelength carriers (unless a large drift field is present), and P/N cells are more efficient than the N/P variety.

Figure 48 shows that the efficiency is fairly constant as a function of junction depth when the surface losses are low. Under these conditions highly efficient GaAs solar cells can be made by epitaxial growth of the top region as well as by diffusion; the epitaxial layer can be several microns thick as long as the electron diffusion length in this layer is 3 µm or higher. Such a condition is difficult (but not impossible, see Table 3) to achieve in GaAs.

3. $Ga_{1-x}Al_xAs$-GaAs

Reducing the junction depth and incorporating a drift field in the diffused region have proven to be difficult for GaAs. An alternative method for overcoming the surface recombination problem is to grow a thin, transparent alloy layer of $Ga_{1-x}Al_xAs$ on the surface of the GaAs junction. The $Ga_{1-x}Al_xAs$ matches

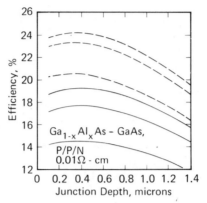

FIG. 49. *Calculated inherent efficiencies of* $pGa_{1-x}Al_xAs$-$pGaAs$-$nGaAs$ *solar cells for 2* $Ga_{1-x}Al_xAs$ *thicknesses (D). In each set of curves, the bottom represents AM0, the middle AM1, and the top AM2. Same conditions as in Fig. 14, Table 5. No fields,* S_{back} = ∞. *—D = 10 μm; --D = 0.1 μm.*

the lattice of GaAs very closely, so that the interface between the two (which is now the "surface" of the GaAs p-n junction) has very few electronic states and a correspondingly low re- combination velocity. At the same time, the bandgap of the $Ga_{1-x}Al_xAs$ is high enough (2.1 indirect, 2.6 direct) to allow most (≳60%) of the light to get through to the underlying GaAs, and the doping level of the alloy layer is high enough to help in reducing the series resistance, allowing the p-region of the GaAs to be doped more lightly for a higher lifetime and better collection efficiency.

Excellent open circuit voltages and fill factors have been obtained in $pGa_{1-x}Al_xAs$-$pGaAs$-$nGaAs$ devices with $Ga_{1-x}Al_xAs$ thicknesses of 2-10 μm [9], but the short circuit currents have been only 20-22 mA/cm^2 for AM0. Analyses of the spectral responses of devices which have such thick $Ga_{1-x}Al_xAs$ layers show that surface recombination losses are not the cause of the low I_{sc}'s in these devices; in fact the measurements indi- cate that surface recombination losses have been eliminated ($S_{interface}$ ≤ 10^4 cm/sec). For these $Ga_{1-x}Al_xAs$ thicknesses, the short circuit current and efficiency are limited by the absorption of high energy light in the $Ga_{1-x}Al_xAs$, which pre- vents the light from reaching the GaAs where carriers can be generated and collected. To maximize the efficiency, the thickness of the $Ga_{1-x}Al_xAs$ must be reduced and the Al content increased (to raise the direct bandgap value), both of which allow more light to penetrate to the GaAs.

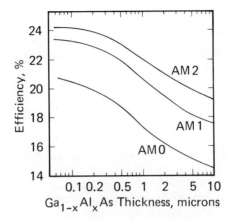

FIG. 50. *Calculated inherent efficiencies of pGa$_{1-x}$Al$_x$As-pGaAs-nGaAs solar cells as a function of Ga$_{1-x}$Al$_x$As thickness. Same conditions as Figs. 14, 49, and Table 5. No fields, S$_{back}$ = ∞, x$_j$ = 0.4 µm.*

The AM0, AM1, and AM2 efficiencies of pGa$_{1-x}$Al$_x$As-pGaAs-nGaAs solar cells are shown in Fig. 49 as a function of junction depth (width of the pGaAs region) for two different Ga$_{1-x}$Al$_x$As thicknesses. The recombination velocities are taken as 10^6 cm/sec at the device surface and 10^4 cm/sec at the interface, while the P/N GaAs portion of the cell is assumed to have the parameters of Table 5. For Ga$_{1-x}$Al$_x$As layers of a few thousand angstroms thickness, the efficiencies are over 20, 23, and 24% for AM0, AM1, and AM2, respectively. It would be very difficult to reach these efficiencies without the Ga$_{1-x}$Al$_x$As layer; comparable AM0 values in straight GaAs devices with recombination velocities of 10^6 cm/sec can only be reached with very narrow junctions and with aiding drift fields, as shown in Fig. 48.

The effect of the Ga$_{1-x}$Al$_x$As thickness on the efficiency is seen more clearly in Fig. 50, for devices with 0.4 µm junction depths and the parameters of Table 5. The increase of efficiency with decreasing thickness can be attributed in part to the greater penetration of light to the GaAs as mentioned above, and in part, to the collection of some of the carriers generated in the Ga$_{1-x}$Al$_x$As [45]. The Ga$_{1-x}$Al$_x$As thickness could theoretically be reduced to as low as 100 Å; the energy barrier in the conduction band (Fig. 13) will continue to prevent photogenerated electrons from entering the alloy layer. (Below this thickness electrons will begin to recombine at the surface after tunneling through the Ga$_{1-x}$Al$_x$As, and the efficiency will drop.) However, the benefits of reducing the

thickness to less than a few thousand angstroms are relatively
small, and it will probably be better to keep the layer around
3000-5000 Å thick to minimize series resistance.

4. SERIES AND SHUNT RESISTANCE LOSSES

    The actual efficiencies of practical devices are less than
the calculated values in Figs. 43 to 50 due to the reflection
of incident light, the portion of the surface masked by the
metallic contacts, and the loss of power in the series and shunt
resistances. The reflection of light is minimized by applying
one or two layer antireflective coatings; the reflection averages
around 9% over the visible spectrum for a one-layer coating and
2-3% for a two-layer coating. The reflection varies strongly
with wavelength for a single layer coating, and the thickness
· of the coating is adjusted to obtain a minimum at around 5500-
6000 Å where the peak in the sun's power occurs (see Fig. 19).
    The contact grid is an important part of the cell; for a
given sheet resistivity of the diffused region, the contact
grid design plays a large part in determining the series resistance. The more the surface is covered with metallic contacts,
however, the less the amount of light is that can get through
to be converted into electric power, so that a trade off exists
between minimizing the series resistance by adding many contact
"fingers" (Fig. 1) and minimizing the amount of surface area
covered by the Ohmic contact. In most devices today, about 5-
7.5% of the surface is masked by the contacts; this can probably
be reduced to 3-4%, but not much less than this. The contact
loss can be included in the efficiency calculations through the
active and total device areas A_a and A_t.
    No solar cell can be made without having some nonzero value
of series resistance. The resistance of the diffused and base
regions and the contact resistances to both those regions all
add to R_s, but the largest factor by far in good devices is the
sheet resistance of the thin, diffused layer. The effects of
series resistance on the AM0 efficiencies of 1 ohm-cm N/P Si
devices and 0.01 ohm-cm P/N GaAs devices are demonstrated in
Fig. 51, where the devices are assumed to have a 1 cm^2 area.
A resistance of 1 ohm for a 1 cm^2 device drops the efficiency
by 1% for Si (15% down to 14%) and 0.6% for GaAs (19.4% to
18.8%); the effect of series resistance is slightly less on
GaAs devices than on Si because the photocurrent is less for
GaAs. The efficiency drops very rapidly for resistances of
several ohms or more.
    Figure 51 can be used for 2, 4, 8, etc., cm^2 area devices
by dividing the abscissa by the area in cm^2; for instance, the

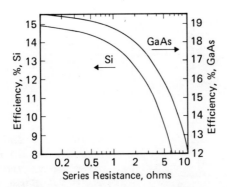

FIG. 51.   The effect of series resistance on the AM0 efficiencies of Si and GaAs solar cells.   Si N/P cell, 1 ohm-cm, parameters of Table 4, x_j = 0.4 μm.   GaAs P/N cell, 0.01 ohm-cm,
parameters of Table 5 with x_j = 0.5 μm. S_{top} = 10^5 cm/sec,
S_{back} = ∞. R_{sh} = ∞.   No drift fields.

drop of 1% in efficiency due to R_s occurs for 1 ohm for a 1 cm^2
sample, 0.5 ohm for a 2 cm^2 sample, 0.25 ohm for a 4 cm^2 sample,
and so forth.

The effects of shunt resistance on the AM0 efficiencies
of Si and GaAs solar cells are demonstrated in Fig. 52.   Ideally,
the shunt resistance should be as large as possible, but a variety of problems can lead to low values of R_{sh}, including edge
leakage and the effects of contact sintering (which can allow
metal "bridges" to form in the vicinity of cracks and scratches
in the device [23]).   Shunt resistance effects are not very
significant unless the value of the resistance becomes less
than 1000 ohm for a GaAs device or 500 ohm for a Si device,
both of 1 cm^2 area.   The effect of R_{sh} is stronger in GaAs devices than in Si ones (just the opposite of the series resistance effects) due to the higher output voltage of GaAs.

A given value of shunt resistance has less effect on large
devices than small ones, since the output voltage is independent
of device area (the current loss equals the output voltage divided by the shunt resistance).   For a 4 cm^2 device, for example,
the shunt resistance could be roughly three times smaller than
the values mentioned above for a 1 cm^2 device before the same
drop in efficiency takes place.   (Of course, increasing the
device area increases the amount of edge leakage and the probability of metal bridges being formed, so that the shunt resistance may be lowered, instead of remaining constant, when the
area is increased.)

In general, surface passivation and careful preparation
of devices prevents significant shunt resistance problems, and

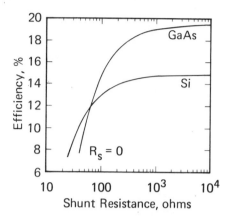

FIG. 52.   The effect of shunt
resistance on the AM0 efficiencies of Si and GaAs solar cells.
$R_s = 0$.   Same conditions as
Fig. 51.

they can often be ignored. On the other hand, series resistance
problems are almost always significant and often reduce the
output by 1 to 1.5 mW per square centimeter of device area.

B.   Measured Efficiencies

     If the measured efficiencies of 1 and 10 ohm-cm Si solar
cells are corrected for reflection, contact loss, and series
resistance loss, the result is in good agreement with the
"inherent" efficiencies of Si devices as described on previous
pages, taking into account the dead layers, high surface recombination velocities, and lower-than-expected base lifetimes
that are often found in actual devices. For Si, the most common type of solar cell is the 10 ohm-cm N/P variety, which has
an AM0 efficiency of about 11.5% (12.8% after correction for
contact area). These devices have uncorrected efficiencies of
14% at AM1 and 16% at AM2. A few 1 and 2 ohm-cm devices have
been made with slightly higher values (as predicted by theory),
while devices made with 0.1 ohm-cm bases have not been as good,
partly because the base lifetime and diffusion length have not
been as high as they should be. The open circuit voltages of
devices with 0.1 ohm-cm base resistivities have been considerably
lower than expected. Both excess tunneling currents (possibly
due to unexpectedly high defect densities in the vicinity of
the depletion region) and effects due to high doping levels
have been suggested as possible causes of the discrepancy. High
defect densities could lead to high depletion region recombination currents as well as to excess tunneling currents.
     Application of the back surface field concept to 1 and 10
ohm-cm devices of 8 to 10 mil thickness should result in improvements by as much as one percent (i.e., 11 to 12% uncorrected)
over devices without the BSF.

The violet cell, with its very good short wavelength response due to the elimination of the dead layer, is generally made with 2 ohm-cm p-type base material [53], and has reached AM0 efficiencies of 14-14.5% uncorrected and 15-15.5% after contact area correction [4]. At AM1 these devices have been up to 18% efficient after correction.

It should be only a matter of time before a violet cell of 8-10 mil thickness with a BSF is made; these will probably be another percent higher than the violet cells made in the past.

Most P/N Si cells have been lower in efficiency than N/P devices, in agreement with Fig. 44. The Li-doped P/N cells developed recently, however, are an exception, because the hole lifetimes and diffusion lengths in Li-doped n-type Si are comparable to the electron lifetimes and diffusion lengths in boron-doped p-type Si. The measured AM0 efficiencies of Li-doped P/N cells have achieved 11.9-12.8% [82,83], which becomes 12.9-13.8% after contact area correction. These high efficiencies, combined with the enhanced radiation tolerance of the Li-doped cells, make them strong contenders to replace the standard N/P cells in satellite applications.

Experimental GaAs solar cells, in spite of their high predicted efficiencies, have always been lower than their expected capability by nearly a factor of 2, almost surely due to much lower lifetimes and diffusion lengths than expected from bulk measurements. Measured efficiencies at AM0 have been about 10% before correction and 11% after correction for contact area [38]. At AM1, the corrected efficiencies have reached 13% [6]. The experimental devices made so far have had fairly deep junctions (≥ 0.5 μm) and relatively poor lifetimes in both regions; the use of high quality starting material and of methods to obtain shallower junctions will probably bring higher measured efficiencies in the near future.

$pGa_{1-x}Al_xAs-pGaAs-nGaAs$ devices have proven to have significantly better efficiencies than conventional GaAs cells, with AM0 values of 13-13.5% [9,84] after contact area correction (12-12.5% before correction). These devices have corrected AM1 efficiencies of 16-17% and corrected AM2 efficiencies of 20-21% [8]. The change in efficiency with air mass value is larger for these devices than for Si or conventional GaAs cells because of the cutoff in response at around 2.5 eV in the $Ga_{1-x}Al_xAs$ devices made so far. Thinner $Ga_{1-x}Al_xAs$ layers should increase the efficiency and decrease the dependence on atmospheric conditions because of the better spectral response at high photon energies.

The theoretical efficiencies of thin film Cu_2S-CdS solar cells have not been calculated here because there is no complete

TABLE 6
V_{OC} and FF (calculated).   300°K

Silicon N/P			Silicon P/N		
ρ	V_{OC}	FF	ρ	V_{OC}	FF
(Ω-μm)	(V)		(Ω-μm)	(V)	
10	.545	.81	10	.525	.80
1	.60	.82	1	.575	.82
0.1	.70	.84	0.1	.63	.83

GaAs P/N			$Ga_{1-x}Al_xAs$-GaAs, P/P/N		
S_{front}	V_{OC}	FF	D	V_{OC}	FF
(cm/sec)	(V)		(μm)	(V)	
10^7	.92	.820	2.5	.945	.82
10^6	.935	.822	0.25	.95	.825
10^5	.95	.825			

quantitative model as yet which can predict all the phenomena
going on in these devices, particularly the events taking place
at the heterojunction interface. Considerable progress has
been made in understanding these devices, however, due princi-
pally to the work of Lindquist, Fahrenbruch, Bube, Böer, Phillips, and others, and it should only be a matter of time before
a satisfactory model is developed. The highest measured AM0
efficiency for thin film devices made in the laboratory is 8.1%,
and AM0 efficiencies of 5.5% are achieved routinely on production line thin film cells [85]. Cu_2S-CdS cells made from single
crystal CdS yield about the same values.

Typical open circuit voltages and fill factors calculated
for Si, GaAs, and $Ga_{1-x}Al_xAs$-GaAs devices are given in Table 6.
These are median values; differences in junction depths, surface
recombination velocities (at the front and back), and lifetimes
result in variations around them. The calculated space charge
region recombination currents are negligible for Si cells if
the parameters of Table 4 are used together with the S-N-S
theory, but the recombination currents can become significant
if the lifetime within the depletion region is lower than expected.

Measured open circuit voltages in Si 1 and 10 ohm-cm cells are close to the calculated ones in Table 6, but measured voltages in 0.1 ohm-cm devices are considerably lower than the 0.7 V predicted by theory, as mentioned above. The fill factors in Si devices are usually 0.75-0.78, slightly lower than the predicted values, due mainly to the series resistance and perhaps slightly to shunt resistances and space charge layer recombination currents.

Measured open circuit voltages and fill factors in GaAs devices have generally been lower than the computed values of Table 6, with V_{OC} from 0.90 to 0.94 V and FF from 0.76 to 0.79. Both the computed and measured space charge layer recombination currents are proportionally much higher for GaAs devices than for Si cells [59], and along with series resistance, these currents are the probable cause of the low fill factors.

The measured open circuit voltages and fill factors of $Ga_{1-x}Al_xAs$-GaAs devices have ranged from 0.94 to 1.0 V and 0.77 to 0.81, respectively, in good agreement with the calculated values of Table 6. The higher measured V_{OC}'s compared to Table 6 might indicate that n_i for GaAs is indeed slightly smaller than 1×10^7 cm^{-3}, as suggested by Sell and Casey [81]. The slightly lower measured fill factors compared to the calculated values are probably the result of series resistance.

Typical open circuit voltages and fill factors measured for fresh thin film Cu_2S-CdS devices are around 0.45-0.50 V and 0.60-0.65, respectively, although these vary somewhat for different preparation conditions and as the device ages.

C. Summary

In the past, the most important figure of merit for a solar cell has been the efficiency at AM0, i.e., in outer space. The prospects of the wide scale use of solar cells for terrestrial use make the efficiency at AM1 (earth's surface) equally important.

The efficiencies at AM0, AM1, and AM2 have been calculated for a wide range of material parameters and for the three models of uniform doping, constant electric fields, and a back surface field. In each case the photocurrents are found by the methods of Chapter 2 and the total dark current by the equations of Chapter 3. The dark current is subtracted from the photocurrent and the maximum value of the voltage-net current product is determined. The result is the "inherent" efficiency of the device, which does not include the losses due to contact area masking, reflection of light from the surface, and series and shunt resistances.

The inherent efficiencies of Si solar cells at AM0 under optimum conditions range from 18% for 10 ohm-cm base material to 21% for 0.1 ohm-cm material. At AM1, the corresponding values are 20-22%. GaAs inherent efficiencies exceed 23% at AM0 and 25% at AM1. The predicted efficiencies are considerably below these values when dead layers, high surface recombination velocities, low base lifetimes, or poor lifetimes within the depletion region are present, leading to 14-15% predicted effi- ciencies for Si and 11-12% for GaAs at AM0, and 16-17% for Si and 13-14% for GaAs at AM1. Reductions in junction depth and the incorporation of electric fields help to overcome these losses, but the best results are obtained if the base lifetime and diffusion length can be improved and the dead layer can be eliminated.

The actual efficiencies of Si and GaAs solar cells are 10-20% less than the inherent efficiencies due to reflection, contact area loss, and series and shunt resistances. The re- flection loss amounts to 6-9% for a single antireflection coating. The upper Ohmic contact grid masks about 5-8% of the surface, reducing the short circuit current accordingly. Series and shunt resistance losses account for the remainder. The series resistance should be less than 0.5 ohms for a 1 cm^2 device (0.25 ohm for a 2 cm^2 cell, etc.), and the shunt resis- tance should be greater than 1000 ohm for a 1 cm^2 device (2000 ohm for a 2 cm^2 cell, etc.) to reduce the resistance losses to acceptably low values.

The present measured efficiencies of Si solar cells are 15% AM0, 18% AM1 for the "violet" cell, and 13.8% AM0, 16% AM1 for Li-doped cells. GaAs cells with $Ga_{1-x}Al_xAs$ layers are 14% efficient at AM0 and 18% at AM1. Cu_2S-CdS devices are 8% effi- cient at AM0 and 10% at AM1.

CHAPTER 5

Thickness

Up to a few years ago, the thickness of a solar cell was not a very important consideration. The short circuit current of a Si device was known [86] to decrease as the base thickness was reduced, and most of the early devices were therefore made with 16-18 mil (400-450 μm) widths to obtain the highest collection efficiency. On the other hand, the power-to-weight ratio and the radiation tolerance to high energy particles both improved with decreasing base thickness, so attempts were made to develop thinner cells with acceptable characteristics, and eventually 10-12 mil (250-300 μm) cells became common.

In recent years, the device thickness has taken on increasing importance. One reason is the advent of the "back surface field" concept and its influence on device behavior. Another is the possibility that thin solar cells can be made cheap enough to have a significant impact on large-scale power generation on earth. These are added to the benefits of high power-to-weight ratios and high radiation tolerance as mentioned above.

When a beam of monochromatic light is incident on a solar cell, some of it is reflected from the front surface, some is absorbed in the bulk of the cell, and some is lost by complete transmission through the device. If the absorption coefficient is high, as it is for short wavelengths, nearly all the light is absorbed near the front surface and the transmission loss is negligible. For long wavelengths where α is low, a high percentage of the light may be lost. The back contact may reflect a portion of this back into the device, but for planar back contacts made in the conventional manner this portion is small.

When the thickness of a solar cell is reduced, the behavior is affected in two ways. First, the loss due to transmitted light increases (the loss is greater for indirect gap materials than for direct gap ones); second, the influence of the back contact becomes greater. The collection efficiency may be

93

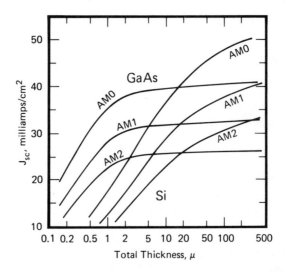

FIG. 53. *Idealized short circuit current densities for 100% collection efficiency as a function of the solar cell thickness.*

reduced and the dark current increased due to excess recombination of photogenerated and dark injected carriers at the back surface. This influence only becomes important when the minority carrier diffusion length in the base is comparable to or larger than the device thickness [5,43,44,87].

Both the ribbon Si technology and the thin film polycrystalline Si technology are aimed at influencing the terrestrial use of solar cells to economically provide large amounts of electrical power. The ribbon technology involves the growth of thin (4-8 mil) Si single crystal ribbons an inch or so wide and arbitrarily long; ideally, the ribbons are produced in a manner suitable for further processing (diffusion, contacting, etc.) without cutting, lapping, or polishing. Efficiencies of 10% at AM1 have already been demonstrated by Mlavsky and co-workers at Tycho [88]. Solar cells made from thin Si films (about 10 μm) grown on cheap metallic or glass substrates hold the promise of being ultimately cheaper than Si ribbon devices if grain boundary effects can be minimized, but at the present time only 1-2% efficient devices have been made [89].

In this chapter the effect of thickness on the short circuit current, open circuit voltage, fill factor, and inherent efficiency of single crystal Si and GaAs p-n junction solar cells for single light passes will be discussed, including the influence of an Ohmic back contact or back surface field on the device characteristics (the width W_{p+} of the BSF region is assumed to

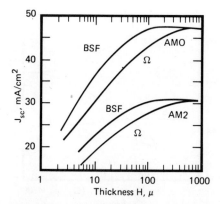

*FIG. 54. Calculated short circuit current densities of
10 ohm-cm Si N/P solar cells, for both an Ohmic back contact
and a BSF back contact. Parameters of Table 4, with S_{front} =
10^5 cm/sec, x_j = 0.2 μm. Drift field present in top region.*

be negligible). The chapter ends with some speculation about
the role of grain boundaries in limiting device efficiencies
and some words on the effect of thickness on CdS devices.

A.   Single Crystal

1.   SILICON

When sunlight is incident on the surface of a solar cell,
the long wavelength portions of the spectrum may be progressively
lost as the device is made thinner. Indirect gap materials such
as Si with their low absorption constants over much of the spectrum suffer much more from the effects of thickness than direct
gap materials such as GaAs, where most of the carriers are generated close to the surface. If it is assumed that light makes
a single pass through the device and that every photon absorbed
creates a hole-electron pair which is collected, then an idealized short circuit current can be calculated as shown in Fig. 53.
Silicon devices begin to lose current when the thickness becomes
less than 500 μm, and GaAs devices when the thickness becomes
less than 3 μm.

In real cases where normal bulk and surface recombination
losses take place, both the magnitude of the photocurrent and
its thickness dependence are reduced. The thickness dependence
does not become appreciable until the thickness becomes less
that about 2 base diffusion lengths (232, 164, and 52 μm for
10, 1, and 0.1 ohm-cm Si, respectively). Figure 54 shows the

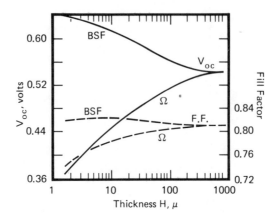

FIG. 55.   Calculated open circuit voltages and fill factors of 10 ohm-cm Si solar cells, for same conditions as Fig. 54.

J_{sc} as a function of thickness for 10 ohm-cm N/P devices using the parameters of Table 4 and the equations of Chapter 2. When the back surface is covered by an Ohmic contact ($S_n = \infty$), the current density decreases as H is reduced below 400 μm. The back surface field ($S_n = 0$) however, with its carrier confinement and reduced recombination loss, maintains the current density at a relatively constant level for thicknesses over 100 μm.[a]

The benefits of a BSF were first noted experimentally through the higher open circuit voltages obtained from such devices compared to conventional devices of the same resistivity Figure 55 shows the open circuit voltage and fill factor calculated for the 10 ohm-cm device of Fig. 54. The most striking benefit of the BSF is in the V_{oc}. For 100 μm devices, as in Si ribbon cells, more than 50 mV are gained by a BSF compared to an Ohmic back contact, and the gain becomes greater as the thickness is reduced further. A beneficial effect of the BSF is also obtained on the fill factor.

Actually, the effects of thickness on the V_{oc} should be separated from the effects of the back contact. As the thickness of the base is reduced, the amount of *bulk* recombination that can take place is reduced proportionally, which tends to

[a]An Ohmic contact acts as a perfect sink for minority carriers and a perfect source for majority carriers. A BSF contact acts as a perfect source for majority carriers but blocks minority carriers.

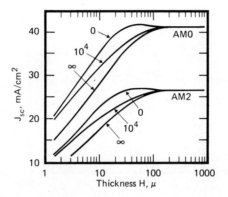

FIG. 56. Calculated short circuit current densities of
0.1 ohm-cm Si solar cells, for both an Ohmic back contact
(S_{back} = ∞) and a BSF back contact (S_{back} = 0 and 10^4 cm/sec).
Drift field present in top region. Parameters of Table 4,
with S_{front} = 10^5 cm/sec, x_j = 0.2 μm.

lower the dark current and raise V_{oc}. If an Ohmic back contact
is present, however, then a surface of infinite recombination
velocity is brought closer and closer to the junction as the
base thickness is reduced. The resulting increased surface
recombination outweighs the reduced bulk recombination, raises
the dark current, and lowers V_{oc}. On the other hand, if a BSF
is present at the back instead of an Ohmic contact, then back
surface recombination is unimportant and the benefits of reduced
bulk recombination on lowering the dark current and raising the
V_{oc} can be obtained.[b]
    The effect of thickness on 1 and 0.1 ohm-cm devices is
qualitatively similar to the effect on 10 ohm-cm devices, but
the thickness dependence does not become important until lower
thicknesses are reached because of the smaller diffusion lengths.
Figure 56 shows the currents obtained for various thicknesses
and back surface conditions for a 0.1 ohm-cm N/P device. A zero
recombination velocity at the back would result in some enhance-
ment of the current compared to an Ohmic contact. The BSF may
not yield a zero SRV at the back for this resistivity, however,
because of the reduced value of ψ_p; calculations for a recombination velocity of 10^4 cm/sec are shown in Fig. 56 therefore to

[b]It has been suggested that part of the added V_{oc} might be due
to the small built-in barrier ψ_p (Fig. 7) and the fact that some
of this barrier voltage might appear in the output [90].

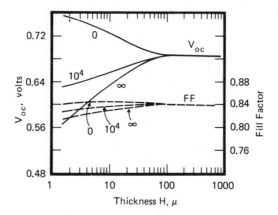

*FIG. 57.   Calculated open circuit voltages and FF of 0.1 ohm-cm
Si N/P solar cells, for same conditions as Fig. 56.*

simulate a partial carrier confinement. Figure 57 shows the
open circuit voltage and fill factor for the device of Fig. 56.
The improvements in these two quantities due to a low SRV at
the back are not as large as in the 10 ohm-cm device, but are
noticeable nonetheless.

Figures 58 and 59 show the efficiencies calculated for 10,
1, and 0.1 ohm-cm N/P devices at AM0 and AM1 as a function of
device thickness for both zero surface recombination velocity
(BSF) at the back and for an Ohmic back contact. When the BSF
condition exists, the efficiency peaks at thicknesses between
2 and 4 mil. Ribbon Si devices with thicknesses of around
4 mil or so would be near the optimum, with efficiencies of
almost 18% at AM0 and 20% at AM1 (20.6% at AM2). Even 10 μm
thick single crystal cells are capable of high efficiencies,
16% at AM0 and 18% at AM1. A zero surface recombination velocity at the back is beneficial over the entire thickness range
from 1 μm up to about twice the base diffusion length, as seen
in Figs. 58 and 59. If an Ohmic back contact has been made
at the back surface rather than a BSF, the efficiencies are
5 to 6% lower (i.e., 12% instead of 18%).

Much of the recent experimental work on thin Si solar
cells has centered around the BSF effect. Mandelkorn and
Lamneck [5] reported considerably higher V_{oc}'s (0.58 V) for
4 mil 10 ohm-cm BSF cells than for 4 mil conventional 10 ohm-cm
devices ($V_{oc} = 0.50$ V), and improvements in short circuit current and fill factor were seen as well. The AM0 efficiencies
of BSF cells changed very little with decreasing thicknesses
compared to conventional devices, in agreement with Fig. 58;
η dropped from 12 to 11.3% going from 12 to 4 mil thicknesses,

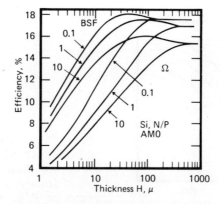

FIG. 58. *Inherent AM0 efficiencies of Si N/P devices for different base resistivities and thicknesses. Parameters of Table 4, with S_{front} = 10^5 cm/sec, x_j = 0.2 μm. Drift field present in top region. Ohmic back contact and BSF-type contact.*

while conventional cells dropped from 11 to 8.3% for the same thickness change.

More recently, V_{OC}'s of up to 0.6 V have been reported for 10 ohm-cm BSF cells [43,90].

Godlewski *et al.* [43] have pointed out that the experimental open circuit voltages of BSF cells made so far have tended to be constant as a function of thickness, while the assumption of a zero surface recombination velocity at the back predicts an increasing V_{OC} with decreasing thickness. The explanation for this discrepancy might be that the effective SRV at the back is not zero but an intermediate value such as 10^3-10^4 cm/sec [the value of S in Eqs. (52) and (53)], implying that some degree of minority carrier leakage across the barrier ψ_p (Fig. 7) takes place. In this case a constant value of V_{OC} as a function of thickness can be predicted. The higher the barrier height ψ_p and the more defect-free the back diffusion is, the lower the effective surface recombination velocity is likely to be. Another possible explanation involves the component added to V_{OC} by ψ_p; this component should be largely independent of thickness.

Iles and Zemmrich [87] have described experimental thin devices (10-2 mil) incorporating both a thin diffused top region as in the violet cell and a back surface field to prevent recombination at the back contact. Efficiencies of 13.75, 13.4, and 10.7% at AM0 for 10, 4, and 2 mil thicknesses were measured (contact area uncorrected). These very encouraging results lead to high expectations for thin, single crystal ribbon Si devices. Exploration has just begun on devices made from Si thin films (10 μm) on metal, glass, or other foreign substrates;

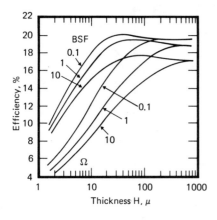

FIG. 59. Inherent AM1 efficiencies of Si N/P devices for
different base resistivities
and thicknesses. Same conditions as Fig. 58.

Chu [89] has reported 1.5% efficiencies at AM1 for thin, polycrystalline Si films on graphite, and expectations are that
this can be brought to 5% in a few years.

Redfield [91] has described the effect of multiple light
passes through Si thin films on the short circuit current of
thin Si solar cells, and has suggested that 2 μm thick films
could have high collection efficiencies for 10 light passes or
more. The multiple light passes would be obtained by a reflec-
tive back surface and total internal reflection at the front.
No mention was made of what the dark current might be in such
thin Si devices, however, and this would have to be taken into
account before the efficiency of thin film multiple pass devices
could be estimated.

2. GALLIUM ARSENIDE

The idealized short circuit currents of GaAs solar cells
as a function of device thickness for a single light pass were
shown in Fig. 53. The decrease in current with thickness is
very slight for thicknesses down to 2-3 μm due to the high absorption coefficient over the entire solar spectrum above
1.4 eV. Even 1 μm thick devices could ideally retain a photocurrent of over 35 mA/cm^2 at AM0 and over 28 mA/cm^2 at AM1.

For actual devices the surface recombination loss can
reduce the current considerably, but the surface recombination
problem can be minimized by making the junction depth small
(as in the Si "violet cell") or by reducing the recombination
velocity at the front (as in the $Ga_{1-x}Al_xAs$-GaAs device).
Figure 60 shows the short circuit current, open circuit voltage,
and efficiency at AM0 for GaAs P/N solar cells. The benefits

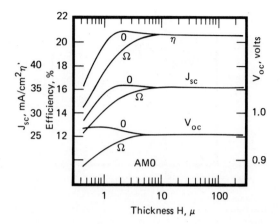

FIG. 60. Inherent AM0 efficiencies, short circuit current densities, and open circuit voltages of 0.01 ohm-cm GaAs P/N solar cells as a function of thickness. The fill factor is 0.82-0.83. Parameters of Table 5, with S_{front} = 10^6 cm/sec, x_j = 0.2 μm. Drift field present in top region. Ohmic back contact and BSF-type contact.

of a low recombination velocity at the back can be clearly seen for all three parameters, and the efficiency peaks for a device thickness between 2 and 3 μm. Even for cells only 1 μm thick, an open circuit voltage of 0.97 V, a photocurrent of 34 mA/cm², and an efficiency of 19.5% are predicted under optimum conditions.

The short circuit currents and efficiencies for the same conditions as in Fig. 60 but at AM1 and AM2 are shown in Fig. 61. Except for the higher efficiencies and lower photocurrents, the behavior for these solar spectra is qualitatively the same as at AM0.

The parameters used for the calculations of Figs. 54-61 are characteristic of relatively good bulk material (Tables 4 and 5). Such good quality material is not often obtained in thin films, particularly if the thin films are grown upon foreign substrates. To demonstrate what might be expected from poor quality material (but still single crystal), the inherent efficiencies at AM0, AM1, and AM2 have been calculated for poor conditions and are shown in Fig. 62. (Silicon: τ_{top} = 3×10^{-9} sec, L_{top} = 0.44 μm, τ_{base} = 0.5 μsec, L_{base} = 16.5 μm, S_{front} = 10^6 cm/sec, S_{back} = ∞. GaAs: τ_{top} = 0.2×10^{-9} sec, L_{top} = 0.57 μm, τ_{base} = 3×10^{-9} sec, L_{base} = 0.95 μm, S_{front} = 10^7 cm/sec, S_{back} = ∞.) The efficiencies are considerably lower than in Figs. 59 and 61, as expected from the lower lifetimes and

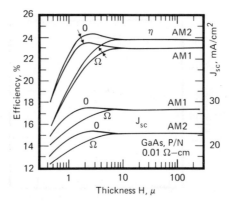

FIG. 61.   *Inherent efficiencies and short circuit current densities at AM1 and AM2 of 0.01 ohm-cm GaAs P/N solar cells as a function of thickness.   Same conditions as Fig. 60.*

diffusion lengths and higher recombination velocities at the front surface.   Nevertheless, the efficiencies are high enough to be adequate for terrestrial use, and the 10% AM1 efficiency for 10 μm thick Si devices and the 9% AM1 efficiency for 1 μm thick GaAs devices offer exciting prospects for the large-scale use of thin film solar cells.   (It should be emphasized that these calculations are for single crystal devices.)

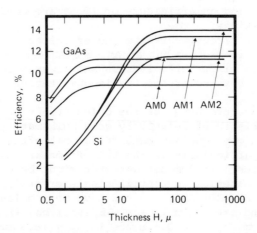

FIG. 62.   *Inherent efficiencies of 0.1 ohm-cm Si N/P solar cells and 0.01 ohm-cm P/N GaAs solar cells for poor quality single crystal material. No drift fields. $S_{back} = \infty$; $x_j = 0.3$ μm.*

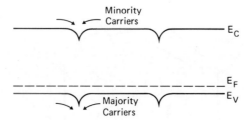

FIG. 63. *Energy band diagram around grain boundaries. The*
boundaries act as "sinks" for minority carriers and barriers
to the movement of majority carriers.

B. Polycrystalline Devices

Single crystal Si, GaAs, or CdS devices can be thinned
down to about 1 mil by a combination of lapping and etching
(lapping, though, introduces lattice damage which lowers the
lifetime and the mobility), and ribbon Si devices can be grown
with an initial thickness of 1-2 mil. For the most part, however, when thin solar cells are desired (<1 mil), they are
fabricated by evaporating, sputtering, or vapor growing a thin
film of material onto a suitable substrate. Under some circumstances the grown films can be single crystal, but they are
more often polycrystalline. Therefore, it becomes important
to understand what the effect of grain boundaries might be on
the internal physics of solar cells.

A thorough analysis of grain boundary effects on both the
photocurrent and dark current in thin solar cells has never
been carried out. A complete analysis of this problem entails
a three-dimensional solution to the diffusion equation with
eight or more boundary conditions. Shockley [92] has given a
partial solution for the simpler case of uniform generation of
minority carriers throughout the volume of a rectangular filament, and used his analysis to define a "filament lifetime"
made up partly of the bulk lifetime in the filament and partly
of surface recombination terms. When the filament dimensions
are small and the surface recombination velocities are high,
the recombination in the filament is dominated by the surface
terms and the "filament lifetime" is much smaller than the
bulk lifetime.

Similar considerations apply to polycrystalline films,
which can be thought of as many filaments connected in parallel
(and sometimes in series). Experimentally, it appears that
grain boundaries act as minority carrier sinks (surfaces of
high recombination velocity) and majority carrier barriers [93],
the worst possible combination (Fig. 63); they reduce the

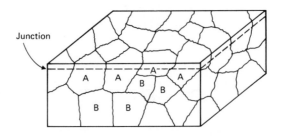

FIG. 64.   Polycrystalline film with random grain orientation.

photocurrent, increase the dark current, decrease the shunt
resistance, and increase the series resistance.   If the grains
are *randomly* oriented, as pictured in Fig. 64, then only the
topmost grain or two (e.g., A in Fig. 64) will be able to contribute to the output; the grains below (e.g., B in Fig. 64)
are effectively isolated from the junction by the grain boundaries above them.   This is the "series" combination of grain
boundaries; the effective lifetime in the film will be very low
and the device will behave poorly.   If the thin film is fibrous
epitaxial, however, as in Fig. 65, then minority carriers within
each "filament" can cross the junction boundary and the whole
layer thickness can contribute to the output.   For this case,
the overall solar cell can be thought of as a parallel combination of filamentary solar cells, each of which act in a normal
manner except that minority carrier recombination can take
place on the "sides" of the filaments as well as at the front
surface and back contact.   This type of film can make a fairly
efficient device under some circumstances, while the randomly
oriented films probably cannot (unless the grain size happens
to be of the order of a diffusion length or more).

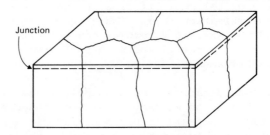

FIG. 65.   Polycrystalline film with fibrous orientation.

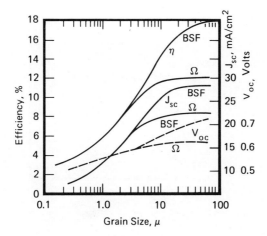

FIG. 66. *AMl inherent efficiencies, current densities, and open circuit voltages of 10 μm thick polycrystalline 0.1 ohm-cm Si N/P solar cells. $S_{front} = 10^5$ cm/sec, $x_j = 0.2$ μm. Starting base diffusion length (large grain sizes) same as in Table 4. Drift field present in top region.*

    Even though the three-dimensional equations for thin film polycrystalline devices have not been solved, it would be beneficial to have an estimate of what the behavior of such devices might be. Such an estimate can be made by constructing a logical argument for the effect of the fibrously oriented grain boundaries on the minority carrier distribution. Consider a solar cell with a thickness H and a grain size of 5 H (e.g., 10 μm thick Si films with a 50 μm grain size). Then most of the minority carriers generated inside a given grain on the base side of the junction will have at least 2 1/2 times as far to travel to a grain boundary as to the junction edge (65% of the carriers are generated within the first 2.5 μm from the surface in a 10 μm thick device). The photogenerated carriers will then have a high probability of being collected, and the device will behave very much as a single crystal solar cell.

    As the grains become smaller, a loss of photocurrent and an increase of dark current due to recombination at the grain boundaries will begin to occur (the effective diffusion length and the effective lifetime for minority carriers in the base begin to decrease). By the time the grain size is down to the device thickness, minority carriers in the base have at most an equal probability of recombining or of being collected, and the effective diffusion length in the base is then at best about equal to the grain size. For grain sizes less than the

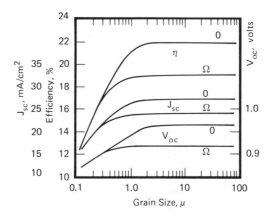

FIG. 67. AM1 inherent efficiencies, current densities, and open circuit voltages of 1 μm thick polycrystalline 0.01 ohm-cm GaAs P/N solar cells. S_{front} = 10^6 cm/sec, x_j = 0.2 μm. Starting base and top region diffusion lengths (large grain sizes) same as in Table 5. Drift field present in top region. Ohmic back contact and BSF-type contact.

device thickness, the diffusion length in the base is less than or equal to the grain size, and the effective lifetime is reduced according to the normal relation $L = \sqrt{D\tau}$.

The philosophy for the effect of grain boundaries on recombination in the top region and depletion region is the same; the grain boundaries may not be important (for fibrous orientation) until the grain size becomes less than about 5-10 times the thickness of these regions. For a 2000 Å junction depth, grain sizes down to around 2 μm should have little effect on the collection from the top region. Grain sizes down to a half-micron or so should have little effect on the collection or carrier recombination within the 500 Å wide depletion region (0.1 ohm-cm base material).

There is some experimental justification for assuming the minority carrier diffusion length equal to the defect size for thin semiconductor layers. Ettenberg [94] studied the effect of dislocation density on the diffusion lengths in GaAs layers grown on GaAs and GaP substrates, and concluded that for high dislocation densities (such that subgrains form in the epitaxial layer) the diffusion length became roughly equal to the average spacing between defects (i.e., the grain size in our model).

Using these optimistic arguments for the effect of grain boundaries on diffusion lengths and lifetimes, the short circuit currents, open circuit voltages, and efficiencies have

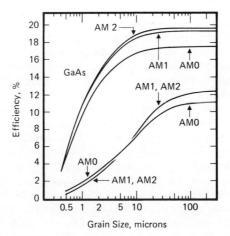

FIG. 68.   Inherent efficiencies of Si and GaAs solar cells for the same conditions as in Figs. 66 and 67 except that the diffusion lengths and lifetimes are determined by the Soclof-Iles analysis [95].   Ohmic back contact.

been calculated at AM1 for 10 μm thick Si devices and 1 μm thick GaAs devices as a function of grain size for fibrously oriented grains, as shown in Figs. 66 and 67.   The Si cell loses efficiency very slowly down to a grain size of 10 μm and more rapidly thereafter.   An efficiency of 10% can theoretically be obtained with grain sizes as low as 3 μm, and 5% for 0.7 μm grains.   The effect of a low recombination velocity at the back is clearly seen for large grains, but becomes negligible for grain sizes below about 2 μm (since for small diffusion lengths, neither the photogenerated nor dark minority carriers "see" the back of the cell).

The GaAs cells of Fig. 67 have higher efficiencies than the Si devices of Fig. 66, and are better than Si in sustaining their efficiencies with decreasing grain size.   The steep absorption coefficient and relatively small diffusion lengths in GaAs make grain boundaries less important than in Si, particularly for very thin (∿1 μm) GaAs devices, and relatively good efficiencies are predicted under optimum conditions even for 1 μm grains, provided the grains are fibrously oriented.   Even random grain polycrystalline GaAs films may still yield acceptable device behavior if the grain sizes are 0.5 μm or more.

The calculations of Figs. 66 and 67 represent optimistic estimates for the effect of grain boundaries on device behavior. Soclof and Iles [95] have made significantly lower estimates of device efficiency as a function of grain size, using a two-dimensional analysis with fibrously oriented grains.   Their

analysis in essence assumes that the effective diffusion lengths
in the base and top regions are reduced to about one-fifth of
the grain size by the grain boundary recombination regardless
of the thickness of these regions. Calculations of the inherent
AMO, AM1, and AM2 efficiencies of 10 μm thick Si cells (0.1 ohm-cm
N/P) and 1 μm thick GaAs cells (0.01 ohm-cm P/N) are shown in
Fig. 68, using the starting parameters of Tables 4 and 5 but
allowing the diffusion lengths and lifetimes to decrease with
grain size according to the Soclof-Iles analysis. The efficiencies predicted by this analysis are substantially lower than
those predicted in Figs. 66 and 67, particularly for grain sizes
less than 10 μm. It is possible that efficiencies somewhere
between the optimistic values of Figs. 66 and 67 and the less
optimistic values of Fig. 68 might be obtained someday, but for
the moment at least, while the quality of such thin films remains
poor (as in Fig. 62) in addition to the grain boundary problem,
efficiencies closer to those of Fig. 68 will more likely be obtained.

It hardly needs to be mentioned that there are severe
practical problems to fabricating solar cells from polycrystalline layers, chief of which is the junction formation. Diffusion tends to proceed faster along grain boundaries and dislocations than in the bulk, and low values of shunt resistance can
result if the diffusion along grain boundaries penetrates through
or almost through to the substrate. It is probably better to
use Schottky barriers or heterojunctions to fabricate cells on
polycrystalline layers, together with some method of reducing
recombination at grain boundaries (see below).

One of the earliest experimental polycrystalline Si devices
was that of Heaps et al. [96] in 1961, who grew 1-2 mil thick
randomly oriented poly films on Si substrates and obtained 0.6-
0.9% efficiencies for "photoflood lamp" light. Since their
junction depth was 2.5 μm and the random grain orientation
limited the active device thickness to 3-4 μm, efficiencies
higher than this would not be expected.

The high interest in thin Si solar cells for terrestrial
uses has lead to some more recent work on thin polycrystalline
films. Chu [89] has deposited 1-2 mil Si films onto graphite
substrates at around 1000°C and obtained 1.5% efficiency at AM1.
Chu, Fang, and others [89,96-98] are attempting to deposit
10 μm polycrystalline films on steel and Al substrates because
of the low cost of these substrates. Iles [99] has suggested
doping the grain boundaries more heavily than the bulk, which
would reverse the "cusp" in the energy bands at the boundary
(Fig. 63) and lower the recombination there. (This will work
provided the diffusion does not penetrate the entire layer and
give high leakage.) No reports of efficiencies from devices

made in these ways have appeared as yet, but experiments are
presently under way and will undoubtedly be reported on soon.
      Even less work has been done on GaAs thin film devices
than on Si ones. Vohl *et al.* [100] have described the characteristics of 15-50 μm thick GaAs films deposited by vapor growth
on Mo or Al substrates. Solar cells of 4.3-4.6% efficiency for
tungsten light were made using either evaporated Pt Schottky
barriers or a Cu$_{1.8}$Se layer to form a heterojunction. Attempted
p-n junctions in these films were less successful because of
rapid diffusion of Zn along the grain boundaries and a resulting
low shunt resistance.

C. Cu$_2$S-CdS

      The behavior of thin film CdS solar cells as a function
of thickness is somewhat more involved than for Si and GaAs
devices. The CdS cells are made by evaporating a CdS film about
1 mil thick onto a transparent plastic substrate, immersing in
a Cu ion plating solution to form a thin Cu$_2$S layer, depositing
a contact grid, and encapsulating in a second transparent plastic
sheet. Depending on the manner in which the back contact is
made, light can be incident on the cell from either the Cu$_2$S
side or the CdS side.
      CdS is a direct bandgap material with high densities of
states in the conduction and valence bands. The absorption
coefficient therefore rises to very high values (10^5 cm^{-1}) for
photon energies even slightly higher than the bandgap [101]
(2.4 eV). For this reason the active region on the CdS side
of the heterojunction is only a micron or so wide.
      For a device with light incident on the CdS side, reducing
the thickness of the CdS layer will have essentially no effect
at all (from a theoretical viewpoint) until the CdS thickness
becomes less than 2-3 μm. At that point more light would be
able to reach the active regions of the device (the depletion
region width plus up to about 1 diffusion length in both the
CdS and Cu$_2$S), and the efficiency would theoretically increase
with decreasing CdS thickness. When the thickness became less
than a diffusion length, however, the efficiency would decrease
because of lost carriers that would otherwise be collected.
      When light is incident on the Cu$_2$S side, reducing the CdS
thickness would again have no appreciable effect until the
thickness became comparable to or less than a diffusion length;
for thinner CdS layers, carriers are lost that might otherwise
be collected and the efficiency decreases.
      This simple reasoning argues that the CdS layer in a
Cu$_2$S-CdS solar cell can be reduced to the order of a micron

without harmful effect, and in a single crystal device this
would be true, but for a thin film cell the situation is a bit
more complicated. The CdS layer is polycrystalline, and the
grain size is probably a function of distance from the CdS-
substrate interface. If the CdS is made thinner, the number
of defects in the active region will probably increase and the
collection efficiency will deteriorate. In addition, the Cu
ions migrate down the grain boundaries during the Cu plating
and subsequent heating steps, and if the polycrystalline CdS
is too thin, they may migrate to the back contact and cause high
leakage. The CdS layer is usually made 20 μm thick or so to
prevent such problems from arising.

Some of the effects of the Cu_2S thickness have already been
discussed in Chapter 2. This layer is responsible for all the
response between 1.1 and 2.4 eV. If light is incident on the
Cu_2S side, then there is an optimum thickness of the Cu_2S of
around 2000 Å. If the Cu_2S is thinner than this, the long wavelength response is lost and the series resistance increases;
if it is appreciably thicker, then carriers are generated too
far from the interface and the short wavelength response suffers.

If light is incident on the CdS side, then the Cu_2S thickness is less important, and it should be made thicker than usual
for lower series resistance and higher long wavelength response,
provided that the longer plating times needed for thicker layers
do not result in Cu ion migration along grain boundaries and
subsequent low shunt resistances.

D. Summary

As the thickness of a solar cell is reduced, losses may
begin to appear due to complete transmission of longer wavelength
light through the device (possibly modified by reflection at the
back surface) and due to increased recombination at the back
Ohmic contact. By comparison with its value at large thicknesses
the maximum possible short circuit current in Si is reduced to
92, 70, and 25% for 100, 10, and 1 μm devices, respectively.
In GaAs the drop in current is not appreciable until the thickness is reduced below 2 μm.

The largest effect of thickness on solar cell behavior is
due to the back electrical contact. An Ohmic back contact reduces the photocurrent and increases the dark current by acting
as a surface of high recombination velocity. A blocking back
contact (such as a BSF) enhances the photocurrent and lowers
the dark current by preventing recombination at the back surface
Neither effect becomes noticeable, however, unless the device
thickness is less than about 2 diffusion lengths in the base.

The inherent AM1 efficiencies of 100 μm thick single
crystal Si devices for relatively good quality material range
from 15% (10 ohm-cm) to 19% (0.1 ohm-cm) for Ohmic back contacts and 17% (10 ohm-cm) to 20% (0.1 ohm-cm) for blocking back
contacts. 10 μm thick single crystal devices have optimum AM1
efficiencies of 8% (10 ohm-cm) to 12% (0.1 ohm-cm) for Ohmic
back contacts and 15% (10 ohm-cm) to 18% (0.1 ohm-cm) for blocking back contacts. The behavior of GaAs devices changes much
less with thickness compared to Si cells because of the much
smaller diffusion lengths in GaAs. The inherent AM1 efficiencies
of 1 μm thick GaAs devices for relatively good quality material
range from 19% for an Ohmic back contact to 21.5% for a blocking
back contact. The efficiencies of both Si and GaAs single crystal cells are reduced strongly if the lifetime in the material
is poor and the front surface recombination velocity is high.

In polycrystalline solar cells, grain boundaries present
the biggest problem because of the severe loss of minority
carriers that can occur at the boundaries. The effect of the
grains becomes important in 10 μm thick Si devices when the
grain size becomes less than 50 μm, and in 1 μm GaAs devices
when the grain size becomes less than 3 μm. Reasonably good
AM1 efficiencies (7% or more) are predicted for 10 μm thick Si
devices if the grain sizes exceed 10 μm and the grains are
fibrously oriented. AM1 efficiencies of 10% or higher are predicted for 1 μm thick GaAs devices for grain sizes of 1 μm or
more.

The effect of thickness on thin film Cu_2S-CdS solar cells
is very small. The CdS thickness could theoretically be reduced
to several microns before any appreciable change took place.
In practice, however, the quality of the Cu_2S and CdS are probably a function of the CdS thickness, and this might result in
poor behavior if the CdS thickness were reduced to less than
5 μm.

CHAPTER 6

Other Solar Cell Devices

Although the great majority of solar cells are made with
p-n junctions, there are several other types that could exhibit
unique advantages of one kind or another, including Schottky
barriers, heterojunctions, vertical multijunction devices, and
grating cells. The Schottky barrier cell is very simple and
economical to fabricate, and has improved spectral response at
short wavelengths, but the expected efficiencies may be somewhat
lower than conventional cells because of lower open circuit
voltages. Heterojunction solar cells can also have enhanced
short wavelength response, and are potentially as efficient as
conventional cells under optimum conditions. Heterojunction
and Schottky barrier cells could be very important for terrestrial applications because of potentially low cost and because
they do not necessarily entail diffusion processes, which can
be detrimental to polycrystalline devices. Vertical multijunction solar cells are potentially high in efficiency and
radiation tolerance, and they could become important for high
intensity applications. Grating cells could theoretically have
both higher short circuit currents and higher open circuit
voltages than conventional devices, and are relatively simple
to make. Each of these types of solar cells will be discussed
in turn.

A. Schottky Barrier Cells

If a metal is brought into contact with a clean surface
of a semiconductor material, a readjustment of charge takes
place in order to establish thermal equilibrium and under most
conditions an energy band bending occurs at the interface much
as in a p-n junction. If the metal is thin enough to be partially transparent to light (while still maintaining an acceptably low sheet resistivity), then some of the incident light
can penetrate to the semiconductor and a photocurrent will

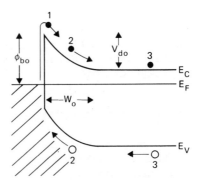

FIG. 69. Energy band diagram
of a Schottky barrier on an
n-type semiconductor.

result. Figure 69 shows the most simple form of an n-type
Schottky barrier device and defines some of its important
parameters: the depletion width W_0, the barrier height ϕ_{b0},
and the diffusion voltage V_{d0}. There are three photoeffects
that can take place. Light can be absorbed in the metal and
excite electrons over the barrier into the semiconductor (1
in Fig. 69); this effect is commonly used to measure the barrier height ϕ_{b0}. Long wavelength light is usually absorbed
deep in the semiconductor, creating hole-electron pairs just
as in a p-n junction; the holes must then diffuse to the junction edge to be collected (3 in Fig. 69). Shorter wavelength
light entering the semiconductor is absorbed partly in the
bulk and partly in the depletion region (2 in Fig. 69), and
very short wavelength light is absorbed entirely in the depletion region. The high electric field in this depletion region
"sweeps" the photogenerated carriers away before they can
recombine at interface states, resulting in good collection
of these carriers, in contrast to a p-n junction where a low
lifetime in the top region and a high surface recombination
velocity can seriously lower the response at short wavelengths.
     The excitation of carriers from the metal into the semi-
conductor is a much smaller effect [102] (by a factor of 100
or more) than the band-to-band excitation mechanisms (2 and 3
in Fig. 69), and can be neglected compared to these for sunlight applications. As far as the photocurrent generation
and collection are concerned therefore, the Schottky barrier
solar cell can be thought of as a p-n junction with a zero
junction depth but with an attenuating metal coating at its
surface. The high field in the depletion region of the Schottky
cell serves the same function as the drift field in the diffused
region of a normal p-n cell in overcoming surface losses.
     In a Schottky barrier solar cell, just as in a p-n junc-
tion cell, the photocurrent passing through a load causes the
device to be forward biased, and a dark current flows in the

opposite direction to the photocurrent. A V-I characteristic
is obtained under illumination which appears qualitatively
the same as in Fig. 2, although the dark current in the Schottky
barrier is very different in nature from the dark current in
a p-n junction.

1. SPECTRAL RESPONSE AND PHOTOCURRENT

The two major contributions to the spectral response and
to the photocurrent come from the depletion region and from
the bulk (the "base" of the Schottky barrier solar cell). The
collection from the depletion region is qualitatively the same
as in a p-n junction. It is assumed that the high field in
the depletion region sweeps carriers out before they can recombine, leading to a current for monochromatic light equal to

$$J_{dr} = qT(\lambda)F(\lambda)(1-\exp(-\alpha W)) \qquad (75)$$

where $T(\lambda)$ is the transmission of the metal film into the
underlying semiconductor, $F(\lambda)$ is the incident photon flux,
α is the absorption coefficient, and W is the width of the
depletion region, given by [103]

$$W = \sqrt{(2\varepsilon_s/qN_d)(V_d-V_j-(kT/q))} \qquad (76)$$

where ε_s is the static dielectric constant. (The kT/q term
is due to the small density of mobile carriers within the space
charge region.) The reflection of light from the metal surface
is accounted for in $T(\lambda)$. The photocurrent expressed by (75)
is similar to that expressed by (20) except that the transmission of light through the metal [$T(\lambda)$] replaces the transmission
($\exp(-\alpha x_j)$) of light through the top region of the p-n junction
and the reflection of light from the surface of the p-n device.
The collection from the base of the Schottky barrier cell
is qualitatively the same as from the base of a p-n junction
cell, and the equations and derivations of Chapter 2 apply
with only the transmission factor as a modification. The photocurrent due to holes collected from the n-type base region is
then

$$J_p = \frac{qF\alpha L_p}{(\alpha^2 L_p^2-1)} \, T \, \exp(-\alpha W) \left[\alpha L_p \right.$$

equation continues

$$-\frac{(SL_p/D_p)\left[\cosh(H'/L_p)-\exp(-\alpha H')\right]+\sinh(H'/L_p)+\alpha L_p\,\exp(-\alpha H')}{(SL_p/D_p)\,\sinh(H'/L_p)+\cosh(H'/L_p)}$$

$$(77)$$

where S is the recombination velocity at the back contact and
H' is the thickness of the device minus the width of the depletion region: H' = H-W.

If the back contact is Ohmic, (77) simplifies to

$$J_p = \frac{qF\alpha L_p}{(\alpha^2 L_p^2-1)}\,T\,\exp(-\alpha W)\left[\alpha L_p-\frac{\cosh(H'/L_p)-\exp(-\alpha H')}{\sinh(H'/L_p)}\right] \qquad (78)$$

and if the device thickness is much greater than the diffusion
length H' >> L_p, (77) simplifies to

$$J_p = \left[qF\alpha L_p/(\alpha L_p+1)\right]T\,\exp(-\alpha W). \qquad (79)$$

The total photocurrent is found by adding (75) to (77), (78),
or (79) as the conditions warrant. (If a drift field is present
in the base, (31) can be used, setting $x_j = 0$, multiplying by
$T(\lambda)$, and changing the subscripts from n to p.)

It should be noted that Eqs. (75) and (77) for the photocurrent do not explicitly take into account the nature of the
barrier. The derivations are based on the assumption of a
high field in the depletion region which sweeps photocarriers
out of this region regardless of interface states and which
reduces the excess minority carrier density at the depletion
region edge to a negligible value for short circuit conditions.
If interface effects become important, the photocurrent may
be reduced. In particular, the photocurrent may be reduced
if there is an interfacial layer [104] such as an oxide or
another insulator more than 30 or 40 Å thick between the metal
and the semiconductor. For thinner oxide layers than this,
tunneling can take place readily and the photocurrent will not
be seriously affected by the oxide layer. For somewhat thicker
oxide layers, the photocurrent may still not be seriously affected if the layer is conducting rather than insulating.

It has been suggested [105] that an inversion layer adjacent to the interface can lower the response to short wavelengths
by causing electrons generated very near the surface to diffuse
toward the surface along with the drift of photogenerated holes,
even though the drift field would ordinarily cause electrons
to be accelerated *away* from the surface. This theory, however,
only applies for very low light levels [105] ($\Delta n,\Delta p \ll n_0,p_0$
where n_0 and p_0 are the equilibrium carrier densities) and
should have no effect on Schottky barrier cells with incident

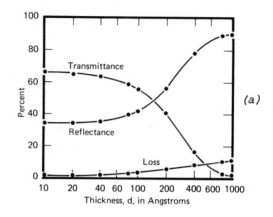

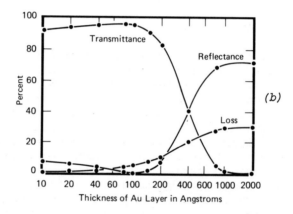

FIG. 70.   *Transmission, reflection, and absorption of 6328 Å*
light through Au films on Si substrates: (a) Au film alone;
(b) with added antireflection coating.   (After Schneider [106],
reprinted with permission from the Bell System Tech. J., Copy-
right 1966, The American Telephone and Telegraph Company.)

intensities greater than about 0.01 sun.   The image potential
discussed in the next section is capable in theory of reducing
the response to short wavelengths, but for any doping level
greater than 10^{15} cm^{-3}, the width of the image force region
is less than 50 Å and photogenerated carriers should not be
seriously affected by it.

The internal spectral response of a Schottky barrier
photodetector is found from

$$SR(\lambda) = [J_{dr}(\lambda) + J_p(\lambda)]/qF(\lambda)T(\lambda),$$ (80)

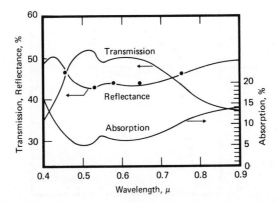

FIG. 71. Transmission, reflection, and absorption of light for 75 Å Au films on GaAs. No AR coating; ●— experimental points. (After Stirn and Yeh [107], courtesy of the IEEE.)

and the photocurrent under sunlight or another type of illumination is found by integration

$$J_{photo} = q \int_0^\infty F(\lambda)T(\lambda)SR(\lambda) \, d\lambda. \tag{81}$$

Equations (75) and (77) for J_{dr} and J_p include the factor $T(\lambda)$ for transmission of light through the metal, but if an antireflective coating is applied to reduce reflection from the metal surface, $T(\lambda)$ represents the transmission through the combined layers [106]. This transmission factor varies from metal to metal and is a strong function of the metal thickness. The transmission of 6328 Å light through thin Au films into a Si substrate as a function of Au film thickness is shown in Fig. 70, as computed by Schneider [106]. If no antireflection coating is applied, the transmission is 55-65% for films less than 100 Å thick. The addition of an antireflection coating increases the transmission to over 90% (Fig. 70b). The transmission of light through a 75 Å Au film into a GaAs substrate as a function of wavelength is shown in Fig. 71, as computed by Stirn and Yeh [107]; an antireflection coating should bring the transmission to above 90% over most of the visible region.

Baertsch and Richardson [108] have given expressions for the reflection and transmission of light through a metal into a dielectric material; these expressions can be used to calculate the transmission through various metal films into various

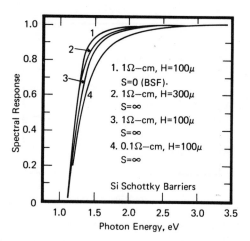

FIG. 72.   *Internal spectral response of Schottky barriers on n-type Si for various device thicknesses and back contact conditions. Parameters of Table 4. Transmission through the metal is assumed to be unity. 1: 1 ohm-cm; H = 100 µm; S = 0(BSF). 2: 1 ohm-cm; H = 300 µm; S = ∞. 3: 1 ohm-cm; H = 100 µm; S = ∞. 4: 0.1 ohm-cm; H = 100 µm; S = ∞.*

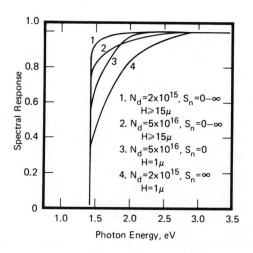

FIG. 73.   *Internal spectral responses of Schottky barriers on n-type GaAs for various thicknesses and back contact conditions. 1: $N_d = 2 \times 10^{15}$; $S_n = 0 - \infty$; $H \geq 15$ µm. 2: $N_d = 5 \times 10^{16}$; $S_n = 0 - \infty$; $H \geq 15$ µm. 3: $N_d = 5 \times 10^{16}$; $S_n = 0$; $H = 1$ µm. 4: $N_d = 2 \times 10^{15}$; $S_n = \infty$; $H = 1$ µm.*

TABLE 7
Short Circuit Current, Devices of Figs. 72 and 73

Material	ρ	H	S(back)	J_{AM0}	J_{AM1}
	(ohm-cm)	(μm)	(cm/sec)	(mA/cm^2)	(mA/cm^2)
Si	1.0	300	∞	44.6	35.4
	1.0	100	∞	42.7	33.9
	1.0	100	BSF	45.2	35.9
	0.1	100	∞	39.8	31.5
GaAs	N_d				
	2×10^{15}	>15	0-∞	38.2	30.7
	5×10^{16}	>15	0-∞	36.2	29.0
	5×10^{16}	1	∞	30.1	23.1
	5×10^{16}	1	0	35.0	27.9

semiconductor materials, provided the optical constants for the metal films are known. Schneider [106] has derived expressions for the transmission of light through a metal film into a dielectric material when an antireflective coating has been applied to the surface of the metal.

The internal spectral response of 1.0 and 0.1 ohm-cm n-type Si Schottky barriers (5×10^{15} and 8.5×10^{16} cm^{-3}, respectively) and of 0.5 and 0.035 ohm-cm (2×10^{15} and 5×10^{16} cm^{-3}, respectively) GaAs Schottky barriers are shown in Figs. 72 and 73, where the transmission $T(\lambda)$ through the metal film is assumed to be unity. These responses were calculated from (75), (77), and (80), assuming a surface recombination velocity at the interface of 10^7 cm/sec and no drift fields outside the depletion region. The high energy (short wavelength) response is seen to approach unity, as a consequence of absorption and carrier generation inside the depletion region where the high field sweeps carriers out with minimum loss. The low energy response is determined by the conditions in the base. A BSF enhances the response (just as it does in a p-n junction) provided the device thickness does not exceed about twice the diffusion length.

The external spectral responses of these same devices with various metals and antireflective coatings on the surface are found by multiplying the internal response of Figs. 72 or 73 by the metal transmission factor $T(\lambda)$.

The short circuit photocurrents of these cells, as calculated from (81) for various conditions, are given in Table 7.

If the transmission factor is high, as it would be for thin
metal films with antireflective coatings, then the photocurrent
is higher for the Schottky cell than it would be in a p-n
junction cell of the same material with the same device thickness, base resistivities, and front and back recombination
velocities. The presence of the depletion region right at the
semiconductor surface in the Schottky cell does indeed go a
long way in overcoming low lifetimes and high recombination
velocities near the surface. On the other hand, p-n junction
cells under the best of conditions (no dead layer, drift field
present, small junction depth) have about the same calculated
photocurrents as the best Schottky cells, and the photocurrents
for the Schottky devices in practice will be reduced to some
degree by the presence of the metal film.

2.   ELECTRICAL BEHAVIOR AND EFFICIENCY

    Although the spectral response and photocurrent do not
depend very strongly on the barrier height, the dark current
depends very much on this height. Figure 69 showed an idealized version of a metal-semiconductor contact without an image
potential or an interfacial layer. The barrier height ϕ_{b0} in
the absence of interface states would be determined by the
difference between the metal work function and the semiconductor electron affinity, but for most semiconductors, including
Si and GaAs, a large density of interface states can effectively "pin" the Fermi level in the semiconductor near the
valence band, making the barrier height roughly 2/3 of the
bandgap for n-type Schottky barriers and $(1/3)E_g$ for p-type
barriers. Measured barrier heights for various metals on Si
and GaAs are given in Table 8. The barrier heights for the
two metals Au and Pt are nearly the same in Si as in GaAs, but
otherwise the heights are larger for GaAs devices as expected
from the larger bandgap.
    The interface state densities in Schottky barriers and
consequently the barrier heights appear to be strongly influenced by the nature of the semiconductor surface [103]; ϕ_{b0}
tends to be larger on cleaved surfaces than on chemically
etched ones [109], and depends on the crystalline orientation
as well for polar semiconductors such as GaAs [109]. Differences in surface preparation are usually responsible for the
range of values reported by different workers for barrier
heights of a given metal on a given semiconductor. In particular, barrier heights tend to be higher when an interfacial
layer such as an oxide is present.
    The actual barrier profile in a metal-semiconductor

TABLE 8
Barrier Heights to Si and GaAs[a]

Metal	ϕ_{b0} (V)	Metal	ϕ_{b0} (V)
Si (n-type)		Si (p-type)	
Au	0.80	Au	0.35
Ag	0.56-0.79	Ag	0.55
Al	0.50-0.77	Al	0.58
Cr	0.58	Cu	0.46
Ni	0.67-0.70	Ni	0.51
Mo	0.58	Pb	0.56
Pt	0.90		
PtSi	0.85		
W	0.66		
GaAs (n-type)			
Au	0.90	Cu	0.82
Ag	0.88	Pt	0.86
Al	0.80	W	0.80
Be	0.81		

[a]After Milnes and Feucht [60]; Sze [103]; and Smith and Rhoderick [112].

contact differs from the ideal version of Fig. 69 due to an interfacial layer, an image potential, or both. The interfacial layer is the result of oxidation or other contamination of the material surface prior to depositing the metal. If care is taken, this layer will not be over a few tens of angstroms in thickness, and it can be eliminated entirely by either cleaning the surface just before deposition (e.g., sputter etching) or by firing the barrier metal into the material (as in PtSi). The image potential on the other hand is a fundamental phenomenon that can be changed in magnitude but not eliminated. It is the result of the attractive force experienced by a charge carrier in the vicinity of a metal surface due to an "image charge" of opposite sign induced in the metal. This attractive force reduces the barrier height by an amount known as the image potential $\Delta\phi$:

$$\Delta\phi = \left(\frac{q^3 N_d (V_{d0} - V_j - kT/q)}{8\pi^2 \varepsilon_d^2 \varepsilon_s} \right)^{1/4} , \tag{82}$$

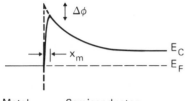

FIG. 74. *Energy band diagram of a Schottky barrier on an n-type semiconductor including the image potential* $\Delta\phi$.

where ε_d is the high frequency dielectric constant and ε_s the low frequency dielectric constant ($\varepsilon_d \approx \varepsilon_s$ for Si and GaAs). In this equation, V_{d0} is the diffusion voltage that would exist if there were no image potential present (Fig. 69).

The barrier height and diffusion voltage are reduced by the image potential according to

$$\phi_b = \phi_{b0} - \Delta\phi, \tag{83}$$

$$V_d = \phi_{b0} - \Delta\phi - (E_C - E_F) = V_{d0} - \Delta\phi. \tag{84}$$

The image potential increases with increasing doping level and decreases with increasing forward bias, and it results in a rounding of the barrier potential profile near the interface, as shown in Fig. 74. The barrier has a maximum value a short distance x_m away from the interface; this distance is about 50 Å for 10^{15} cm^{-3} doped material and decreases to about 15 Å for 10^{17} cm^{-3} material.

The dark current in the forward bias direction of a Schottky barrier is determined mostly by thermionic emission of majority carriers from the semiconductor into the metal [103], for doping levels less than 10^{17} cm^{-3}. The Schottky barrier current can be written as [103]

$$J = A^{**}T^2 \exp(-q\phi_b/kT) [\exp(qV_j/kT) - 1] \tag{85}$$

where A^{**} is the Richardson constant A^* modified by optical phonon scattering, quantum mechanical reflection, and tunneling of carriers at the metal-semiconductor interface. A^* is given by

$$A^* = 4\pi q m^* k^2 / h^3 \tag{86}$$

where m^* is the effective mass tensor for the relevant energy bands in the semiconductor and h is Planck's constant. The dark current (85) is fundamentally different from the forward bias dark current in a p-n junction. In the junction device,

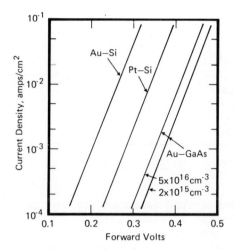

FIG. 75. *Dark current densities of Au-Si and Pt-Si Schottky barriers for 1 ohm-cm Si, and of Au-GaAs devices for two doping levels. Au-Si: $J_s = 5 \times 10^{-7}$ A/cm^2; Pt-Si: $A^{**} = 110$ A/cm^2($^\circ$K)^2; Au-GaAs: $A^{**} = 4.4$ A/cm^2($^\circ$K)^2.*

the current is determined by the rate at which minority carriers can diffuse and drift away from the junction edge after being injected from the opposite side; a BSF reduces the dark current by reducing the rate of this carrier removal. In the Schottky barrier, the dark current is determined by the rate at which carriers are emitted from the semiconductor into the metal, and it is a majority carrier current that is not affected by a BSF on the semiconductor.

This fundamental difference between the two current mechanisms shows up in a very important way; in a p-n junction device, the dark current at a given voltage *decreases* strongly with *increasing* doping level in the base, but in the Schottky barrier, the dark current *increases* (slightly) or remains about the same for increasing doping levels up to around 10^{17} cm^{-3}, and increases very strongly (because of tunneling) above 10^{17} cm^{-3}. Therefore, the open circuit voltage is smaller for high doping levels than for low doping levels under otherwise equal conditions. The slight variations in dark current with doping level arise from the image potential, which affects ϕ_b in Eq. (85), and from the optical phonon scattering and quantum mechanical reflection, which affect A^{**} [103]. Andrews and Lepselter [110] have shown that the value of A^{**} for moderately doped Si varies somewhat with electric field but has an average value of around 110 A/cm^2($^\circ$K)^2 for n-type Si and 32 A/cm^2($^\circ$K)^2 for

TABLE 9
Dark Current Parameters, Au-Si, 300°K[a]

N_d (cm^{-3})	J_s (A/cm^2)	n
1×10^{15}	5×10^{-7}	1.0
5×10^{15}	4.5×10^{-7}	1.0
1×10^{16}	4.5×10^{-7}	1.005
5×10^{16}	5.5×10^{7}	1.015

[a]After Chang and Sze [108].

for p-type. For n-type GaAs, A** is around 4.4 A/cm^2(°K)^2 for moderate doping levels [111].

   The calculated forward bias dark currents for Au-Si, Pt-Si, and Au-GaAs devices without interfacial layers are shown in Fig. 75 for 1 ohm-cm n-type Si and for two GaAs doping levels. The difference in the two Si curves arises from the difference in barrier heights between Au-Si and Pt-Si (Table 8); the barrier height enters exponentially into the dark current determination, as given in Eq. (85). The difference in the two GaAs curves arises from the difference in image potential at the two doping levels (17 mV at 2×10^{15} cm^{-3} and 39 mV at 5×10^{16} cm^{-3}).

   The Schottky barrier dark current is often written as [113]

$$J = J_s [\exp(qV/nkT) - 1] \qquad (87)$$

where n is the slope of the $\ln J$-V curve

$$n = (q/kT)(\partial V / \partial \ln J) \qquad (88)$$

$$\doteq [1 + (\partial \Delta \phi / \partial V) + (kT/q)(\partial(\ln A^{**})/\partial V)]^{-1}. \qquad (89)$$

For most Si and GaAs Schottky barriers made with carefully cleaned surfaces, n is in the range of 1.01-1.03 due to the variation of $\Delta \phi$ and A** with voltage and doping level. Table 9 lists the values of J_s and n for Au-Si devices as computed by Chang and Sze [113]; the values of n are very low for the moderate doping levels used for Schottky solar cells. On the other hand, minority carrier currents (hole injection from the metal and space charge layer recombination), tunneling currents, and interfacial layers can result in considerably higher values of n [104,109,111,113], with values of 2 or more observed in some cases [114].

TABLE 10
Performance of Si and GaAs Schottky Barriers,
300°K, n = 1-1.02

Dev	ρ	H	S(back)	V_{oc}	FF	η_{AM0}	η_{AM1}
	(ohm-cm)	(μm)	(cm/sec)	(V)		(%)	(%)
Au-Si	1	300	∞	0.30	0.72	7.2	7.8
Pt-Si	1	300	∞	0.38	0.76	9.6	10.5
Au-Si	1	100	∞	0.30	0.72	6.9	7.4
Pt-Si	1	100	∞	0.38	0.75	9.1	10.0
Au-Si	1	100	BSF	0.30	0.72	7.2	7.8
Pt-Si	1	100	BSF	0.38	0.76	9.7	10.6
	N_d						
Au-GaAs	2×10^{15}	>15	$0-\infty$	0.47	0.79	10.3	11.6
Au-GaAs	5×10^{16}	>15	$0-\infty$	0.47	0.79	9.9	11.1
Au-GaAs	5×10^{16}	1	∞	0.46	0.79	8.1	8.7
Au-GaAs	5×10^{16}	1	0	0.46	0.79	9.6	10.6

The open circuit voltages, fill factors, and efficiencies for the Schottky barrier cells of Table 7 are listed in Table 10, assuming 100% metal transmission, no series or shunt resistance losses, and single crystal material with bulk lifetimes (the series resistance of most metal films greater than 50 Å thick can be made acceptably low by using a fine pattern contact grid as in the violet cell). Pt-Si devices are more efficient than Au-Si because of the larger barrier heights. GaAs devices are only slightly more efficient than Si devices because the barrier heights are only slightly higher on GaAs than they are on Si (Table 8).

The calculated efficiencies for Schottky barrier solar cells on Si and GaAs are lower than for p-n junctions of these materials at the same doping levels because of the low open circuit voltages obtained from the Schottky barrier cells. The low V_{oc}'s in turn are a result of the dark current mechanism (85) in Schottky barriers and the relatively low values of ϕ_b; if barrier heights equal to the bandgap could be obtained, the "limit conversion efficiencies" of Schottky barrier cells would be about the same as for p-n junction cells [115], i.e., about 22% for Si and 25% for GaAs. As long as the barrier heights remain low, however, as in Table 8 (measured on Si and GaAs with carefully cleaned surfaces) [103], the output voltages will be low. The presence of an interfacial layer

can apparently result in increased barrier heights [104,109],
which would lead to lower dark currents and higher V_{OC}'s, but
the interfacial layer may also reduce the photocurrent by adding an extra energy barrier that photogenerated carriers must
tunnel through.

It has been suggested [107,114] that high values of n in
Eq. (87) can also lead to low dark currents, even without large
barrier heights, because the function $\exp(qV/nkT)$ increases
more slowly with increasing V when n is high. Anderson et $al.$
[116] have used a value of n = 2 to predict open circuit voltages of 0.53 V and efficiencies of 12-16% at AM1 for Cr Schottky
barriers on p-type Si, assuming that J_S in (87) is unchanged
by the high n-value. However, if the higher values of n are
due to excess dark currents, it is obvious they cannot lead to
higher open circuit voltages; the value of J_S will be increased
in this case and lower V_{OC}'s will be obtained. If the higher
values of n are due to interfacial oxides, on the other hand,
then the barrier height is increased, the value of J_S is decreased, and higher V_{OC}'s will be obtained. A trade-off could
exist between the higher open circuit voltage and the lower
short circuit current as a result of the interfacial oxide
layer. Experimentally, the gain in V_{OC} seems to outweigh any
decrease in J_{ph} and improved efficiencies are obtained by adding a thin interfacial layer.

The measured efficiencies of Schottky barrier solar cells
have been less than 10%. Anderson et $al.$ [114,116] have obtained efficiencies under AM1-AM2 sunlight of 8.1-9.5% using
Cr Schottky barriers on 2 ohm-cm p-type Si, with open circuit
voltages of 0.5-0.53 V and short circuit currents of 26 mA/cm^2.
Stirn and Yeh [107] have measured open circuit voltages of
0.53 and 1.0 V for Au-GaAs and Au-GaAs$_{0.6}$P$_{0.4}$ devices, respectively, and obtained a short circuit current density of over
18 mA/cm^2 under AM0 illumination (with no antireflective coating) for Au-GaAs Schottky barriers, leading to expected AM0
efficiencies of 10% after AR coating application. The spectral
response of the Au-GaAs device was high at short wavelengths,
confirming the original premise of improved collection of carriers generated very near the semiconductor surface.

Schottky barrier solar cells could be highly useful for
terrestrial applications where a slightly lower predicted
efficiency compared to p-n junction cells would be far out-
weighed by their inherent simplicity and expected lower cost.
They could be particularly applicable to polycrystalline Si
and GaAs solar cells where normal p-n junction diffusion processes may be difficult because of the presence of grain
boundaries.

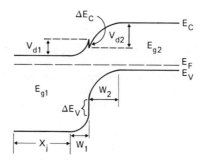

FIG. 76. Energy band diagram of a typical N/P heterojunction in thermal equilibrium. (When voltage is applied, the barriers become $V_{d2}-V_{j2}$ in material 2 and $V_{d1}-V_{j1}$ in material 1.)

B. Heterojunctions

Heterojunction solar cells have many similarities and a few differences to Schottky barrier cells. The most important similarity is that short wavelength photons can be absorbed within or very near the depletion region of the device under most circumstances, leading to good high photon energy response. The most important difference is that the open circuit voltage can be quite high as in a p-n junction, without the need for carefully controlled interfacial layers.

The energy band diagram of a typical heterojunction between two single crystal materials is shown in Fig. 76. Light of energy less than E_{g1} but greater than E_{g2} will pass through the first material (which acts as a "window") and become absorbed by the second material, and carriers created within the depletion region and within a diffusion length of the junction edge will be collected exactly as in a p-n homojunction cell. Light of energy greater than E_{g1} will be absorbed in material 1, and carriers generated within a diffusion length of the junction edge or within the depletion region of this material will also be collected. The advantage that the heterojunction can have over most normal p-n junctions is in the short wavelength response; if E_{g1} is large, high energy photons will be absorbed inside the depletion region of material 2 where the carrier collection should be very efficient. If material 1 is also thick in addition to being high in bandgap, the device should have lower series resistance and higher radiation tolerance than a p-n junction made entirely of material 2.

1.   SPECTRAL RESPONSE AND PHOTOCURRENT

The major contribution to the spectral response and photocurrent comes from the base material, with a smaller contribution from the top material and from the two depletion regions. If a N/P heterojunction is assumed, with a neutral base region (no drift field), then the photocurrent from the *base* can be found from (7), (9), (16), and (17) after taking into account the attenuation by the top layer

$$J_n(\lambda) = \frac{qF(\lambda)\ \exp[-\alpha_1(x_j+W_1)]\ \exp(-\alpha_2 W_2)\alpha_2 L_{n2}(1-R)}{(\alpha_2^2 L_{n2}^2-1)}$$

$$\times \left[\alpha_2 L_{n2} - \frac{\dfrac{S_n L_{n2}}{D_{n2}}\left(\cosh\dfrac{H'}{L_{n2}}-\exp(-\alpha H')\right)+\sinh\dfrac{H'}{L_{n2}}+\alpha_2 L_{n2}\ \exp(-\alpha H')}{\dfrac{S_n L_{n2}}{D_{n2}}\sinh\dfrac{H'}{L_{n2}}+\cosh\dfrac{H'}{L_{n2}}}\right] \tag{90}$$

where S_n is the recombination velocity at the back surface, $F(\lambda)$ is the intensity of monochromatic light incident on the surface of the top layer, α_1 and α_2 are the absorption coefficients of the two materials, W_1 and W_2 are the depletion widths on each side of the interface, and H' is the width of the neutral base region, $H' = H-(x_j+W_1+W_2)$. The reflection of light from the interface due to the difference in refractive indices has been ignored, since this will generally be from 3 to 4% or less. This reflection can be included in (90) by replacing $F(\lambda)\ \exp[-\alpha_1(x_j+W_1)]\ \exp(-\alpha_2 W_2)(1-R)$ by the more involved transmission factor as done by Milnes and Feucht [60]. The depletion widths are determined by the relative doping levels and dielectric constants of the two materials

$$W_1 = [(2\varepsilon_1/qN_1)[\varepsilon_2 N_2/(\varepsilon_1 N_1+\varepsilon_2 N_2)]V_d]^{1/2}, \tag{91}$$

$$W_2 = [(2\varepsilon_2/qN_2)[\varepsilon_1 N_1/(\varepsilon_1 N_1+\varepsilon_2 N_2)]V_d]^{1/2}. \tag{92}$$

The photocurrent at a given wavelength from the *top layer* can be found from (6), (8), (12), and (13)

$$J_p(\lambda) = \frac{qF(\lambda)\alpha_1 L_{p1}}{(\alpha_1^2 L_{p1}^2-1)}(1-R)$$

equation continues

$$\times \left[\frac{\left(\dfrac{S_p L_{p1}}{D_{p1}} + \alpha_1 L_{p1} \right) - \exp(-\alpha_1 x_j) \left(\dfrac{S_p L_{p1}}{D_{p1}} \cosh \dfrac{x_j}{L_{p1}} + \sinh \dfrac{x_j}{L_{p1}} \right)}{\dfrac{S_p L_{p1}}{D_{p1}} \sinh \dfrac{x_j}{L_{p1}} + \cosh \dfrac{x_j}{L_{p1}}} \right.$$

$$\left. - \alpha_1 L_{p1} \exp(-\alpha_1 x_j) \right] \tag{93}$$

where S_p is the surface recombination velocity at the surface of material 1.

The photocurrents from the *depletion regions* are given by

$$J_{W_1}(\lambda) = qF(1-R) \exp(-\alpha_1 x_j) [1-\exp(-\alpha_1 W_1)], \tag{94}$$

$$J_{W_2}(\lambda) = qF(1-R) \exp[-\alpha_1(x_j+W_1)] [1-\exp(-\alpha_2 W_2)], \tag{95}$$

where reflection from the interface has again been ignored in (95).

In order to obtain these expressions, it has been assumed that the excess minority carrier densities at the edges of the depletion region are reduced to zero by the electric field in the depletion region [boundary conditions (13) and (16)]. This is a reasonable assumption provided that the conduction band discontinuity ΔE_C is small ($<kT/q$) in a N/P heterojunction or that the valence band discontinuity ΔE_V is small in a P/N hetero-junction; otherwise, minority carriers on the small gap side of the junction may be impeded from flowing across the junction and the photocurrent will be reduced (for example, electron movement from the base to the top layer is impeded slightly by ΔE_C in Fig. 76). Another complicating factor arises from the presence of interface states and defects caused by lattice mismatch; if the effective lifetime within and around the depletion region becomes very short, photogenerated electrons and holes may quickly recombine instead of being collected, and the photocurrent will again be reduced. The most promising heterojunctions for solar energy conversion are those between pairs of materials with small ΔE_C or ΔE_V, good lattice match, and good thermal expansion match, and in addition, a reliable growth method for fabricating heterojunctions with a maximum of interface perfection is required.

Several heterojunction pairs with good lattice match have been discussed by Milnes and Feucht [60]. These include GaP-Si, ZnS-Si, ZnSe-GaAs, AlAs-GaAs, and $Ga_{1-x}Al_xAs$-GaAs, among others. For GaP-Si and ZnSe-GaAs, ΔE_C is relatively small and N/P solar cells can be made. For AlAs-GaAs and $Ga_{1-x}Al_xAs$-GaAs, ΔE_V is small and good P/N devices can be made.

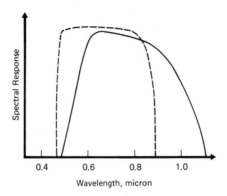

FIG. 77. *Typical spectral responses for heterojunctions with direct and indirect bandgap materials. The solid curve is typical of GaP-Si junctions, and the dashed curve of ZnSe-GaAs junctions.*

Figure 77 shows typical spectral response curves for hypothetical heterojunctions with small energy discontinuities, i.e., Eqs. (90) and (93)-(95) are applicable. The long wavelength (low photon energy) cut-on is determined by the small bandgap material. If the bandgap is indirect, the response will rise gradually with decreasing wavelengths (solid curve), and if the gap is direct, the cut-on will be sharp (dashed curve). The short wavelength cutoff is determined by the larger gap material. Indirect gaps yield gradually decreasing cutoffs (solid curve), while direct bandgaps yield sharp cutoffs (dashed curve).

If the large gap material is thin enough that $x_j < \alpha_1^{-1}$ over most of the solar spectrum or that $x_j < L_1$ (the minority carrier diffusion length), then the short wavelength cutoff will be moved to higher photon energies, since either more light will penetrate to the small gap material or carriers generated in the large gap material will be collected. In either of these two cases, or in the event that E_{g1} is very large (≥ 3.5 eV), the short circuit photocurrent can approach the value that would be obtained in a p-n junction made entirely from the small gap material. For GaP-Si devices made with 1-10 ohm-cm Si and thin GaP layers (<2 μm), for example, the photocurrent theoretically can exceed 40 mA/cm^2 at AM0 and 30 mA/cm^2 at AM1, even with a high recombination velocity at the front and back, and a BSF will have the same effect in improving the photocurrent in the heterojunction device as in a normal Si p-n junction. For ZnSe-GaAs and $Ga_{1-x}Al_xAs$-GaAs devices, the photocurrent could theoretically approach the values calculated for good GaAs p-n junctions, i.e., over 30 mA/cm^2 at AM0 and over 25 mA/cm^2 at AM1.

2. ELECTRICAL BEHAVIOR AND EFFICIENCY

a. *Electrical Characteristics*

While both Schottky barriers and heterojunctions made from a given material can exhibit photocurrents equal to those obtained from p-n junctions in that material, heterojunctions should have larger voltage outputs compared to Schottky barriers, since the heterojunction barrier height is not restricted to some fraction of the bandgap and since the dark current in a heterojunction can be considerably lower (in theory) than in a Schottky barrier. The barrier height V_d in a *homojunction* is given by

$$V_d = E_g - (E_C - E_F) - (E_F - E_V) \tag{96}$$

where $(E_C - E_F)$ and $(E_F - E_V)$ are the differences between the Fermi level and the conduction and valence bands in the n- and p-sides of the junction, respectively. For a *heterojunction*, the barrier height in a N/P device (Fig. 76) is given by

$$V_d = E_{g2} + \Delta E_C - (E_C - E_F) - (E_F - E_V) \tag{97}$$

and in a P/N device is given by

$$V_d = E_{g2} + \Delta E_V - (E_C - E_F) - (E_F - E_V) \tag{98}$$

where E_{g2} is the small bandgap value. The discontinuities ΔE_C and ΔE_V are

$$\Delta E_C = \chi_2 - \chi_1,$$
$$\Delta E_V = (E_{g1} - E_{g2}) - \Delta E_C \tag{99}$$

where χ_1, χ_2 are the electron affinities of the two materials, measured as positive numbers. ΔE_C and ΔE_V can be either positive or negative numbers as determined by (99).

From (97) and (98), the built-in voltage V_d of a heterojunction can be larger than in a homojunction by the amount of the energy discontinuity ΔE_C or ΔE_V if these quantities are positive. At first glance, this seems to hold the promise of higher output from a heterojunction than can be obtained

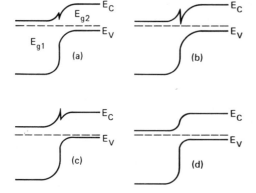

FIG. 78. Energy band
diagrams of N/P hetero-
junctions for several
electron affinity and
doping conditions.

from a homojunction of the small gap material alone; the heterojunction photocurrent could be equal to that of the homojunction
and the barrier height V_d could be larger. Unfortunately, these
two things never occur together because a large barrier height
is accompanied by reduced photocurrent. Figure 78 shows the
energy band diagrams of four N/P heterojunctions with different
values of ΔE_c. In (a), the photocurrent is slightly smaller
than it would be in a homojunction of the small gap material
alone (material 2) and the built-in voltage V_d is about equal
to E_{g2}. In (b), V_d is substantially larger than E_{g2} but the
"notch" in material 2 near the interface hinders transport
across the junction and the photocurrent is greatly reduced.
For both of these cases the large gap material is doped more
heavily than the base, while in (c) the opposite is true. In
Fig. 78c, V_d is larger than E_{g2} but the photocurrent will be
small because of the energy barrier at the interface. In (d),
ΔE_c is a negative quantity $(\chi_2 < \chi_1)$; there are no energy barriers to carrier collection and the photocurrent will be large,
but V_d is smaller than E_{g2} by the magnitude of ΔE_c (Eq. (97)).
     As a general rule, then, the output power from a hetero-
junction solar cell is no larger than it would be from a solar
cell made from either material alone, and the advantages of
heterojunction cells come from eliminating surface recombination and dead layer problems, and from potentially lower series
resistance and higher radiation tolerance. (Heterojunctions
could become important for polycrystalline solar cells also.)
As another general rule, the highest output will be obtained
from devices with small energy discontinuities in the relevant
band, i.e., small ΔE_c for N/P structures and small ΔE_v for P/N
devices (other things being equal).
     There are at least three possible major components to the
forward bias dark current in heterojunctions: injection of

minority carriers from each side of the junction into the other, recombination of holes and electrons within the space charge region, and tunneling.

The injected current is determined by the rate at which minority carriers drift and diffuse away from the junction edge after being injected from the opposite side. The current can be calculated from (37) to (40) just as in a normal p-n junction, but the boundary conditions (41) and (42) are changed due to the energy discontinuities and due to quantum mechanical effects at the interface. A detailed balance of the electron and hole fluxes at the interface is used to obtain the boundary conditions [60]. The current due to electrons injected from the top region into the base is given by

$$J_n = J_0 \exp(qV_j/kT) \tag{100}$$

for $V_j > 2kT/q$, where

$$J_0 = qX_T \frac{D_{n2}}{L_{n2}} \frac{n_{i2}^2}{N_2} \left[\frac{(S_n L_{n2}/D_{n2})\cosh(H'/L_{n2}) + \sinh(H'/L_{n2})}{(S_n L_{n2}/D_{n2})\sinh(H'/L_{n2}) + \cosh(H'/L_{n2})} \right]. \tag{101}$$

X_T is the quantum mechanical transmission factor at the interface (analogous to the quantum mechanical effects in Schottky barriers contained in A^{**}). Since X_T is less than 1, the dark current in the heterojunction due to electron injection into material 2 would theoretically be smaller than in a homojunction of material 2 alone for the same doping level in the base.

For the special case shown in Fig. 78c, ΔE_c is larger than the barrier height $V_{d2} - V_{j2}$ in material 2, where

$$(V_{d1} - V_{j1}) = [\varepsilon_2 N_2/(\varepsilon_1 N_1 + \varepsilon_2 N_2)](V_d - V_j) = K_1(V_d - V_j),$$
$$(V_{d2} - V_{j2}) = [\varepsilon_1 N_1/(\varepsilon_1 N_1 + \varepsilon_2 N_2)](V_d - V_j) = K_2(V_d - V_j). \tag{102}$$

In this case (Fig. 78c), only the fraction of the voltage dropped across material 1 changes the electron distribution, and the injected electron current is then

$$J_n' = J_0 \exp(qK_1 V_j/kT) \tag{103}$$

and is equivalent to J_n except for the different exponential factor. At high enough forward biases, $(V_{d2} - V_{j2})$ will become less than ΔE_c even for the configurations of Fig. 78a and b, and J_n will switch over to J_n' with a change in slope ($\ell n\ J$ versus V) from 1 kT/q to $(1/K_1)kT/q$.

In a homojunction, the second portion of the injected dark current, due to minority carriers injected from the base into the top layer, is usually negligible because of the much lower doping level in the base than in the top layer. In a heterojunction, this portion may be reduced even further by the appropriate energy discontinuity in the valence or conduction band (ΔE_v in a N/P cell and ΔE_c in a P/N device). In the N/P devices of Fig. 78, for example, the energy barrier ΔE_v largely impedes holes from being injected into material 1, and the dark current due to hole injection becomes

$$J_p = qX_T \frac{D_{p1}}{L_{p1}} \frac{n_{i1}^2}{N_1} \left[\frac{(S_p L_{p1}/D_{p1})\cosh(x_j/L_{p1})+\sinh(x_j/L_{p1})}{(S_p L_{p1}/D_{p1})\sinh(x_j/L_{p1})+\cosh(x_j/L_{p1})} \right] \qquad (104)$$

and can usually be neglected because of the low value of n_{i1}^2 (due to the wide bandgap) and because of the high value of N_1 (due to the high doping level).

The second dark current mechanism, due to recombination within the space charge region, is given qualitatively by an expression of the form of (54), although it is not very clear what quantitative effects the energy discontinuities and the interface states will have on this current. For example, for N/P devices like those in Fig. 78, hole injection from the small gap side to the large gap is prevented by ΔE_v, so all the recombination effectively occurs on the small gap side of the interface. Also, if the interface state density is high, the interface recombination velocity is probably also high, and the recombination rate within the depletion region will be much greater than in a p-n homojunction. To calculate the depletion region recombination current comparable to (54), these two factors in heterojunctions would have to be taken into account.

The third current mechanism, tunneling, has proven to be the dominant one in virtually all heterojunctions except those with very small lattice mismatch. Tunneling in p-n homojunctions has already been discussed in Chapter 3, where the current was of the form

$$J_{tun} = K_1 N_t \exp[(4/3\hbar)(m^*\epsilon/N)^{1/2} V_j] \qquad (105)$$

where K_1 is a constant containing the effective mass, built-in voltage V_d, doping level, dielectric constant, and Planck's constant, N_t is the density of available empty states, and N is the doping level in the base. In a heterojunction, the density of available empty states is much larger because of interface states and because of energy states within the

bandgap as a result of cross doping, lattice mismatch, and
thermal expansion mismatch. The tunneling current is also
modified in a heterojunction by quantum mechanical reflection
at the interface and by the fact that only a portion of the
total barrier (V_d-V_j) is normally involved in the tunneling.
For the N/P devices of Fig. 78c or d, for example, tunneling
into interface states in material 1 is given by [61]

$$J_{tun} = K_1 X_T N_{IS} \exp[(4/3\hbar)(m^*\varepsilon_1/N_1)^{1/2} K_1 V_j] \qquad (106)$$

where N_{IS} is the density of available interface states. For
the configurations of Fig. 78a and b, tunneling directly into
interface states does not occur, but tunneling into other
available energy states can take place. When multiple step
tunneling is important, i.e., when the density of energy states
in the forbidden region is so large that carriers can cross
the depletion region by a series of tunneling-recombination
steps, then the current due to tunneling in material 1 takes
the form [61,117]

$$J_{tun} = K_1 X_T N_t \exp[(4/3\hbar)(m^*\varepsilon_1/N_1)^{1/2}(K_1/R^{1/2}) V_j] \qquad (107)$$

where R is the number of steps involved.
     These tunneling currents are very weak functions of
temperature, in contrast to the thermal (injected and space
charge recombination) currents which are strong functions of
temperature. There are other types of tunneling currents that
have temperature dependences midway between (106) or (107) and
the thermal components; for example, electrons may tunnel from
a point part way up the barrier, adding a Boltzmann-type
$\exp(qV/AkT)$ term to the preexponential factor K_1. All of
these possibilities for tunneling currents make it very difficult to predict what the dark current will be for a given
heterojunction. It is safe to say, however, that the more
perfect the junction is (the closer the lattice and thermal
matches and the less the chance for cross doping), the smaller
the tunneling dark current will be and the more efficient the
device will be. Low doping levels in *both* regions (particularly $N_2 < 10^{17}$ cm^{-3} in the base) will also reduce the chance
of tunneling by widening the depletion regions (but, of course,
a low doping level in the top layer might lead to high series
resistance).
     The I-V characteristics of a Cu_2S-CdS heterojunction
device have already been shown in Figs. 40 and 41. The nearly
constant slope of the ℓnJ-V characteristic and the small
change in magnitude with temperature are typical of devices
dominated by tunneling. The great majority of heterojunctions

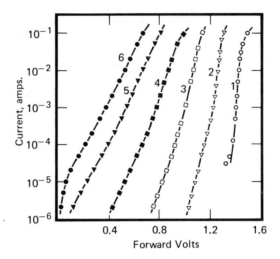

FIG. 79. Current-voltage characteristics of $pGa_{1-x}Al_xAs-nGaAs$ heterojunctions at: (1) 77°K; (2) 180°K; (3) 298°K; (4) 371°K; (5) 433°K; (6) 531°K. (After Alferov et al. [121]; courtesy of the American Institute of Physics.)

have characteristics very similar to these [61]; there are several exceptions, however, including ZnSe-Ge and GaAs-Ge N/P heterojunctions, which have small ΔE_c's, and $Ga_{1-x}Al_xAs$-GaAs devices, which have small ΔE_v's. All three of these devices have good lattice and thermal matches, and all three have demonstrated transistor behavior [118-120], showing that injection *can* be the dominant forward bias current mechanism in some instances. Figure 79 shows the I-V characteristic of a $pGa_{1-x}Al_xAs-nGaAs$ heterojunction with $N_d = 2\times10^{17}$ cm^{-3} in the base and $N_a = 5\times10^{18}$ in the $Ga_{1-x}Al_xAs$ (after Alferov *et al.* [121]). The change in slope and magnitude of the I-V characteristic is typical of heterojunctions where thermal injection dominates.

b. *Efficiency*

A comprehensive calculation of the efficiencies of heterojunction solar cells, including the effects of tunneling currents, recombination at interface states, quantum mechanical effects at the interface, etc., has never been carried out. Several calculations under more idealized conditions have been made, however. Sreedhar *et al.* [122] have computed a type of

TABLE 11
Limit Conversion Behavior, AM0[a]

Heterojunction	V_{OC} (V)	η (%)
GaP-Si, N/P	0.67	24
GaP-InP, N/P	0.81	25
GaP-GaAs, N/P	0.82	21
GaAs-InP, N/P	0.87	27
GaP-Si, P/N	0.69	25
GaP-InP, P/N	0.94	30
GaP-GaAs, P/N	1.05	28
GaAs-InP, P/N	0.93	30

[a]After Sreedhar et al. [122].

"limit conversion efficiency" for various heterojunctions,
where tunneling and space charge layer recombination are neglected, the base is taken as infinitely wide, and 100% collection efficiency is assumed for all photon energies greater
than E_{g2}. Their computed open circuit voltages and conversion
efficiencies are given in Table 11. The predicted efficiencies
are very high, but bear little relation to what might be at-
tained in actual fact.

A more realistic calculation has been performed by Sahai
and Milnes [123], who took space charge layer recombination
currents and reflection of light from the surface into account
and who used equations equivalent to (90)-(95) to compute the
photocurrent. The calculated open circuit voltages and efficiencies for GaP-Si and ZnSe-GaAs heterojunctions are given
in Table 12. The authors made no attempt to optimize the material parameters used in their calculations, and the efficiencies
apply for very particular doping levels, lifetimes, and layer

TABLE 12
Calculated Efficiencies for Heterojunctions
with Close Lattice Match, AM0[a]

Heterojunction	V_{OC} (V)	η (%)
GaP-Si, N/P	0.65	14.3
ZnSe-GaAs, N/P	0.925	15.6

[a]After Sahai and Milnes [123].

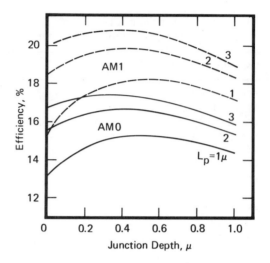

FIG. 80.   Inherent efficiencies at AM0 (solid) and AM1 (dashed) of $pGa_{1-x}Al_xAs$-$pGaAs$-$nGaAs$ solar cells as a function of junction depth, for several base (nGaAs) diffusion lengths. $D = 1$ μm, $S_{back} = \infty$. No fields.

thicknesses, but even so respectable efficiencies are predicted, and better efficiencies would be predicted if all parameters were optimized.

The most successful experimental heterojunction device in terms of efficiency is the $pGa_{1-x}Al_xAs$-$nGaAs$ device of Alferov et al. [7], where 10 μm thick layers of $Ga_{0.4}Al_{0.6}As$ were grown by LPE on 1×10^{17} cm^{-3} GaAs substrates. AM0 efficiencies of 10-11% were measured, with open circuit voltages of 0.95 V and short circuit currents of 15-20 mA/cm^2. The $pGa_{1-x}Al_xAs$-$nGaAs$ solar cell is not quite as efficient as the $pGa_{1-x}Al_xAs$-$pGaAs$-$nGaAs$ structure because the latter collects photogenerated carriers over a larger distance compared to the pure heterojunction [36]. Figure 80 shows the efficiency expected at AM0 and AM1 for pGaAlAs-pGaAs-nGaAs devices as a function of the pGaAs width. The efficiencies can be over 2% higher (15.3% compared to 13%) when the pGaAs region is present compared to the heterojunction case where it is absent, if the base diffusion length is short. When the base diffusion length is large, the pGaAs region has less effect and the heterojunction cell is nearly as efficient as the three layer device.

Several other heterojunction solar cells have been reported also, including GaP-GaAs [124,125], CdS-Si [126], and $CdTe$-Cu_2Te [126a,127]. The efficiencies of the P/N GaP-GaAs devices were around 8% (for unstated conditions), and around 7% for the N/P GaP-GaAs cells under "white light" conditions.

CdS-Si devices, made by evaporation of CdS onto 0.2-3.0 ohm-cm
p-type Si wafers, exhibited about 5% efficiencies for AM1-AM2
sunlight. CdTe-Cu$_2$Te polycrystalline cells had efficiencies
of 6% under AM2 sunlight, while the same cell using single
crystal CdTe substrates exhibited 7-1/2%. This cell is very
similar to the Cu$_2$S-CdS solar cell, which has already been
described at some length in previous chapters.

Fahrenbruch et al. [127a] have reported the experimental
and theoretical efficiencies of several II-VI compound heterojunctions with relatively poor lattice match. The most promising devices were made by vapor growth of p-type CdTe onto
n-type CdS (lattice mismatch 10%), yielding an efficiency in
California sunlight of 4% (theoretical efficiency 17%). The
low experimental efficiencies were attributed to the much
larger dark currents in the finished cells compared to the
theoretical dark currents.

Wagner et al. [127b,127c] have investigated nCdS/pCuInSe$_2$
and nCdS/pInP solar cells. The CuInSe$_2$ devices had a broad
spectral response from 5000 Å to 1.3 μm, and efficiencies of
around 12%. The response of the InP devices was high (70% absolute) from 5000 Å to about 9500 Å, and AM3 (53 mW/cm^2 input)
efficiencies of 12.5% were obtained. These latter devices had
open circuit voltages of 0.63 V, short circuit currents of
15 mA/cm^2, and FF of 0.71.

C. Vertical Multijunction Cells

The vertical multijunction solar cell is an interesting
device concept where light is incident in a direction parallel
to the junction rather than perpendicular to it as in a conventional solar cell. This has some important consequences,
chief of which is that the spectral response (in the ideal
case) becomes very high at all wavelengths, leading to a high
potential efficiency.

The main features of the vertical multijunction solar
cell can be understood with the aid of Figs. 81 and 82. In
Fig. 81a, a conventional N/P solar cell is shown. Light incident on the surface creates hole-electron pairs at a depth
which depends on the absorption coefficient; long wavelength
photons create minority carriers deep in the material which
must diffuse to the junction edge to be collected. The longer
the wavelength is, the greater the distance the generated carrier must travel and the greater the chance of its loss. The
spectral response then increases with decreasing wavelength,
reaches a maximum, and decreases at shorter wavelengths when
most of the carriers are generated in the low lifetime N$^+$ region.

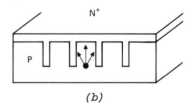

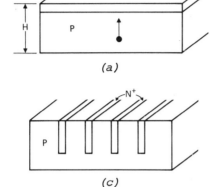

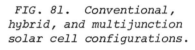

FIG. 81. Conventional,
hybrid, and multijunction
solar cell configurations.

In Fig. 81b, vertical N regions have been added to the conventional structure. Carriers generated deep in the material now have a much shorter distance to travel to reach a junction edge, and if the width between the vertical N regions is less than the diffusion length, the probability of collection is high. The long wavelength spectral response of the structure in Fig. 81b will be high, while the short wavelength response is the same as before due to the losses in the surface N^+ region.

In Fig. 81c, the N^+ sheet covering the surface has been eliminated. Now carriers generated anywhere between the vertical N^+ regions have an equal probability of reaching the junction edge (provided surface recombination is negligible) and the spectral response is uniformly high over the entire wavelength range.

The individual junctions in Fig. 81b and c are connected in parallel by the nature of the way the devices are made (the N^+ regions in Fig. 81c are connected together at one end of the wafer). If the vertical N^+ regions extend all the way through the device, then the junctions can be connected either in parallel or in series as desired. Figure 82 shows several proposed schemes for vertical multijunction devices as proposed by Rahilly, Stella and Gover, Chadda and Wolf, and others [128-134]. In Fig. 82a, the junctions are connected in parallel; the photocurrent will be high while the voltage output will be less than or equal to that of a single junction. In Fig. 82b and c, the junctions are connected in series; the photocurrent will be low but the voltage output will be high, equal in principle to the sum of the voltages of each junction [133].

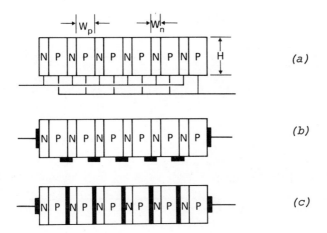

FIG. 82. Vertical multijunction solar cell interconnection schemes.

Regardless of whether a series or parallel interconnection method is used, all the vertical multijunction devices
have the potential of high spectral response over a wide spectral region, provided the surface recombination velocities
are low. Figure 83 shows the spectral response of VMJ devices
of several cell widths (W_p+W_n, Fig. 82) compared to a conventional device for surface recombination velocities of 10^3 cm/
sec, as calculated by Chadda and Wolf [132]. The enhanced
response at long wavelengths is clearly seen, as is the improved response at decreasing wavelength. (The conventional
cell in Fig. 83 is a particularly good one; if a dead layer
were present in the conventional device, the short wavelength
response would be low and the advantage of the VMJ device at
short wavelengths would be more striking.)
    The surface recombination velocity plays a much more
important role for the VMJ device than in conventional solar
cells [128,131,132,134]. If the unit cell width (W_n+W_p) is
large, carriers generated close to the front surface are
strongly influenced by the surface, and recombination velocities exceeding 10^3 cm/sec have a very detrimental effect on
the short wavelength response and the photocurrent [131-132].
By way of contrast, recombination velocities of 10^3 cm/sec
have no appreciable effect on a conventional cell. The recombination velocity at the back surface of a VMJ cell is much
less important to the photocurrent [131] than the recombination
velocity at the front surface (as long as the device thickness
is greater than 50 μm), since most carriers are created closer
to the front surface. Reducing the unit cell width can partially

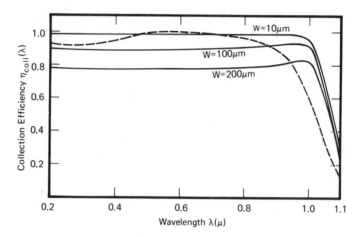

FIG. 83. Internal spectral responses of Si vertical multi-junction solar cells for various unit cell widths (W = W$_n$+W$_p$) and of a conventional Si solar cell. Dashed lines: conventional cell; S$_F$ = 10^3 cm/sec^{-1}. Solid lines: vertical junction cell; S$_F$ = S$_B$ = 10^3 cm/sec. (After Chadda and Wolf [132]; courtesy of the IEEE.)

overcome the effect of front surface recombination by placing generated carriers closer to a junction edge.

Reducing the unit cell width helps to reduce the bulk recombination loss as well as the surface loss, and the bulk recombination loss is essentially negligible as long as L$_i$ $\geq$ 3W$_i$ where W$_i$ is the width of the i region (N or P) and L$_i$ is the minority carrier diffusion length in that region. As long as this condition is fulfilled, the resistivities of the N and P regions can be reduced (to obtain better V$_{oc}$'s) without affecting the spectral response (in a conventional cell reducing the resistivity lowers the long wavelength response and hence the photocurrent).

Lowering the unit cell width is not without its disadvantages, however, since the number of junctions required to make a 1 cm long device is inversely proportional to (W$_n$+W$_p$), and the dark current is proportional to the number of junctions. The number of junctions per cm is therefore one of the important design parameters of the overall device. Rahilly [128] has calculated the performance of parallel VMJ solar cells as a function of the number of junctions N$_j$ and shown that the overall efficiency saturates with increasing N$_j$ due to the opposing influences of increasing dark current and increasing photocurrent. The AMO efficiencies calculated were 16% for N$_j$ $\geq$ 1000 using 1 ohm-cm p-type starting material and 13% for

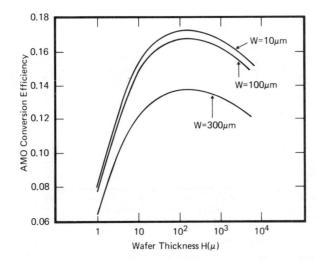

FIG. 84. Inherent AM0 efficiencies as a function of wafer thickness for Si vertical multijunction solar cells with several unit cell widths. (After Chadda and Wolf [132]; courtesy of the IEEE.)

$N_j \geq 2000$ for 10 ohm-cm material. An overall thickness H of 250 μm was assumed.

Chadda and Wolf [131,132] have shown that there is an optimum value of device thickness H on AM0 efficiency. For small thicknesses, some photons are lost by incomplete absorption, and the back surface recombination causes a significant loss as well as the front surface recombination. As H increases, more photons are absorbed and the influence of the back surface decreases, but the dark current through the device increases because of the larger junction area. When H becomes greater than $(1/\alpha)$ for the longest wavelength of importance, then further increases in H simply increase the dark current without improving the photocurrent. Figure 84 shows the AM0 efficiency calculated as a function of the thickness for three unit cell widths. The efficiency peaks at a thickness of ∿100 μm (4 mil), and is over 17% for unit cell widths of 10 μm ($N_j \geq 1000$). Gover and Stella [130] have calculated a peak efficiency of nearly 20% for $(W_n+W_p) = 20$ μm (500 junctions), $S_{front} = 0$, H = 100 μm, and a starting resistivity of 0.1 ohm-cm.

In addition to slightly higher predicted efficiencies for the VMJ cell under optimum conditions compared to conventional solar cells, the VMJ device should be more radiation tolerant, as long as the cell width W_i is much less than L_i

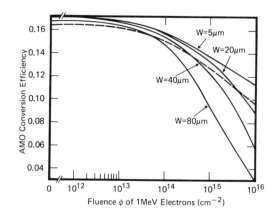

FIG. 85.   Inherent AMO efficiencies as a function of fluence of 1 MeV electrons for a conventional Si cell (broken lines) and for vertical multijunction cells (solid lines) of several unit cell widths.   (After Chadda and Wolf [132]; courtesy of the IEEE.)

(the minority carrier diffusion length) [128,132-134]. This is due to the near-independence of the photocurrent on L_i as long as $L_i \geq 3W_i$. When radiative particles cause a reduction in the diffusion length in the base of a conventional cell, carriers generated deep in the material are lost, but carriers generated near the junction are largely unaffected. In the VMJ device, a reduction in the diffusion length has practically no effect as long as the $L_i \geq 3W_i$ condition prevails; the VMJ device therefore sustains its efficiency to larger fluences than a conventional cell (Fig. 85). On the other hand, when the diffusion length is reduced below this value, the collection efficiency is degraded at *all* wavelengths simultaneously in the VMJ cell, not just in the longer wavelength portion of the spectrum as in conventional devices; the efficiency then degrades at a faster rate with increasing fluence in the VMJ device than in a conventional device. The smaller the unit cell width is (the greater the number of junctions), the more radiation tolerant the device is and the smaller its rate of degration at high fluences, as shown in Fig. 85.

In most of the VMJ structures discussed in the literature, the width of the N region W_n is much less than the width of the P region W_p, and W_p is less than 20 μm or so for optimum device performance. Sater *et al.* [133] have described an alternative where the unit cell is around 10 mil wide, and each unit cell consists of P-N-N$^+$ structures separated by

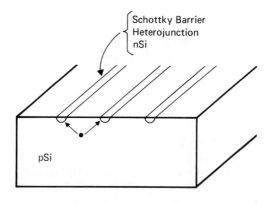

FIG. 86. *Schematic of a grating solar cell.*

layers of Al. From 16 to 96 of these structures are connected in series to form a high voltage, low current solar cell, with outputs of about 6.5 V at 1.1 mA for the 16 cell device, and 36 V at 1 mA for the 96 cell unit. The most important application of this device lies in its capability of operation at very high intensities, since the series resistance is negligible (in the device of Fig. 82c, for instance, carriers have only a very short distance to go to reach an Al contact). One 16-junction device was operated at 150 suns with an AMO efficiency of 6.4%, and exhibited a linear increase of power output with increasing intensity up to this point (no fall-off due to series resistance was observed).

The fabrication processes for VMJ devices are very complex, and the cost may very well outweigh the potential advantages [132] except in a few special cases. Sater *et al.* [133] have described a multiple slice diffusion, alloying, and cutting process that can be used for units up to 100 junctions or so, while Smeltzer *et al.* [134] have described an etching-epitaxial refill method that could be used for VMJ devices having up to several thousand unit cells.

D. Grating Solar Cells

If a semiconductor wafer is contacted by Schottky barriers, heterojunctions, or p-n homojunctions in the form of stripes, as shown in Fig. 86, the resulting device is known as a grating solar cell. Carriers generated by light at some depth beneath the surface must diffuse both vertically and horizontally to the junction region to be collected. If the spacing between

the grids is less than about one minority carrier diffusion
length, the carriers will have a high probability of being
collected, and in particular, if the surface recombination
velocity on the surface of the substrate between the grids is
low, the spectral response to short wavelengths will be considerably better than in conventional diffused structures
with their short lifetime, high recombination velocity "dead"
regions. In addition, the dark current can be lower in the
grating cell than in a conventional cell because of the smaller
junction area, so that both the voltage and current output
from the grating cell can theoretically be higher than from
a conventional cell of the same base doping level, provided
the grating cell is properly designed to prevent series resistance problems.

The two most important design considerations in a grating
cell are the stripe width and the spacing between stripes,
and the recombination velocity at the surfaces between the
stripes [135]; in this respect the grating cell is very similar to the vertical multijunction device. The areas beneath
the stripes will generally have lower spectral responses than
the free areas. If the stripes are Schottky barriers, for
example, the metal will probably be thick enough to block out
all the light, and if the stripes represent p-n junctions,
the short wavelength response will be poor beneath the stripes
because of short lifetimes in the heavily doped regions. The
highest overall response is obtained [135] by minimizing the
stripe width and maximizing the spacing between stripes, as
long as the spacing is kept less than or equal to a diffusion
length and as long as the surface recombination velocity along
the free surface is low.

A back surface field contact to the grating cell will
have the same benefits as it does for conventional cells. If
the grating cell is made with p-n junctions or heterojunctions,
both the open circuit voltage and short circuit current can
be improved by the BSF compared to an Ohmic back contact. If
Schottky barriers are used to fabricate the grating cell, only
the short circuit current will be improved by the BSF.

Figure 87 shows the spectral response of grating cells
made by alloying 8 μm wide Al stripes spaced 95 μm apart into
2 ohm-cm n-type Si substrates [135]. The long wavelength
response is slightly lower than in conventional cells of the
same base resistivity, since carriers generated deep in the
material have a longer distance to travel to reach the junction edge. The short wavelength response is considerably
better than in conventional cells, however, indicating a low
surface recombination velocity ($\leq 10^3$ cm/sec) on the surfaces
of the device between the stripes and a diffusion length

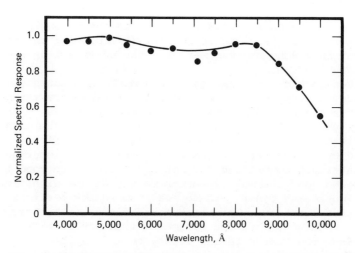

*FIG. 87. Relative (normalized) spectral response of a P/N
Si grating solar cell with 8 μm wide stripes spaced 95 μm
apart. (After Loferski et al. [135]; courtesy of the IEEE.)*

considerably higher than 100 μm. These experimental grating
cells exhibited short circuit current densities slightly higher
than those obtained from conventional N/P cells under the same
spectral source conditions, but their open circuit voltages
were lower than expected, possibly due to excessive leakage
currents around the stripe edges [135]. The prevention of
excess leakage currents (by etching or other techniques) should
result in high open circuit voltages as predicted by theory.

Grating photodiodes have also been made using Au-Si
Schottky barriers [136], but no solar cell measurements were
made on these devices.

E. Summary

Schottky barrier solar cells are very simple and economical
to fabricate. Their main advantage is their response to short
wavelength light, and high short circuit currents can be obtained provided that antireflective coatings are added to minimize the normally high reflection from the surface of the metal.
The dark currents are higher in Schottky barriers than in p-n
junctions with the same base doping level, and this leads to
lower open circuit voltages and efficiencies. If a very thin
interfacial oxide layer is present between the metal and the
semiconductor, however, higher barrier heights and open circuit voltages are obtained, and the efficiency may be improved

as long as the short circuit current is not strongly affected.
A back surface field can be helpful in improving the photocurrent in a Schottky barrier, but has no effect on the dark
current.

Heterojunction solar cells can also have enhanced short
wavelength response, and can have lower series resistances and
better radiation tolerance to low energy particles than conventional p-n junction cells. In order to obtain the high short
circuit currents, open circuit voltages, and efficiencies predicted by theory for optimum conditions, it is important that
the materials comprising the heterojunction have good lattice
match and good thermal expansion match, do not significantly
cross-dope each other, and do not form energy barriers to photocurrent collection. The maximum efficiency of a heterojunction
cannot be higher than the maximum efficiency of a p-n junction
made from the base material alone, but it may be easier to reach
these high efficiencies with the heterojunction due to the absence of a dead layer and due to lower series resistances. A
back surface field is capable of improving the photocurrent and
lowering the dark current, just as in a conventional p-n junction. Experimental heterojunction solar cells have not matched
their high predicted performance, mainly because of excess tunneling currents due to defects at the interface.

Both the heterojunction and Schottky barrier device con-
cepts could be very useful for polycrystalline solar cells
where normal diffusion processes may result in severe grain
boundary problems and/or low shunt resistances.

Vertical multijunction solar cells offer slightly improved
efficiencies, better radiation tolerance, and potentially higher intensity operation compared to conventional solar cells.
The device thickness, unit cell width (number of junctions),
diffusion length, and resistivity are all important design parameters. The surface recombination velocity at the front of the
device is more important than in conventional cells, and should
be 10^3 cm/sec or less for good photocurrent collection. A BSF
can be helpful for both the photocurrent and the dark current.

Grating solar cells have potentially high short circuit
currents, open circuit voltages, and efficiencies, and are very
simple to make. The surface recombination velocity at the front
is very important (as in the VMJ device) and must be low to obtain high photocurrents. The stripe width and the distance between stripes are the most important design parameters, and the
best devices are obtained with minimum stripe widths (properly
contacted to prevent series resistance problems) and maximum
stripe spacings (consistent with keeping this spacing less than
a diffusion length in the base). A BSF can be beneficial in
improving the photocurrent and reducing the dark current.

CHAPTER 7

Radiation Effects

Solar cell behavior under incident particle radiation is of great importance, since the major application for solar cells lies in satellite, space station, and space vehicle power sources. The regions outside the earth's atmosphere are very hostile as far as electronic devices are concerned; high densities of electrons, protons, neutrons, and alpha particles with energies ranging from 1 keV to hundreds of MeV's can play havoc with the minority carrier lifetimes. An idea of the particle distributions trapped by the earth's magnetic field can be seen in Fig. 88, as presented by Hess [137]. In Fig. 88a, the regions around the inner Van Allen belt are shown (shaded area), characterized by a particle density of about 10^4 protons/cm^2-sec with very high energies, from 20 to 200 MeV [138]. The second Van Allen belt, shown in Fig. 88b, is characterized by a high energy electron density of 10^4-10^5/cm^2-sec with energies between 1 and 2 MeV. This belt, and regions beyond it, also contains a high proton density of 10^7-10^8/cm^2-sec with energies in the 1-5 MeV range (Fig. 88c), while a high density (10^6-10^8/cm^2-sec) of low energy electrons and protons (1-100 keV) pervades the entirety of space from the inner Van Allen belt to about 10 earth radii (63,800 km) as shown in Fig. 88d. To this steady-state situation are added the variable fluxes of neutrons, protons, electrons, and alpha particles from variations in the solar wind and the occasional bursts of particles from atmospheric nuclear detonations.

A. Radiation Damage

When radiative particles enter the body of a solar cell, they cause a considerable amount of lattice damage (vacancies and interstitials, vacancy-impurity complexes, defect clusters, and the like). In space, this damage results in a gradual deterioration of performance over a period of time. A single

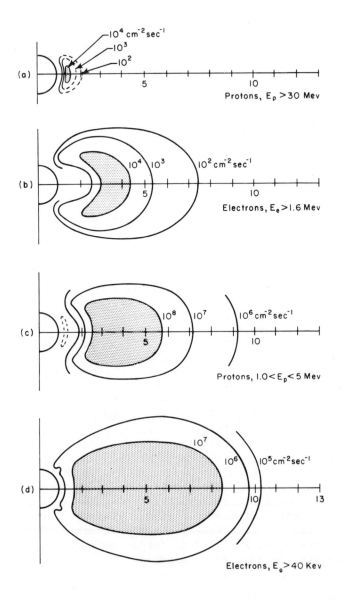

FIG. 88. Radiation zones outside the atmosphere. (Reprinted by permission of the publisher, from "The Radiation Belt and Magnetosphere" by W. N. Hess. Copyright © 1968 by Xerox Corporation. Published by Xerox College Publishing, successor in interest to Blaisdell Publishing Co. (Division of Ginn & Co.))

particle can produce a number of defects which usually act as recombination centers and lower the lifetimes and diffusion lengths in the cells.

Two convenient units of measure have been introduced to describe the effects caused by a given particle at a given energy. These two units are the damage coefficient K and the critical fluence ϕ_c. The damage coefficient describes the rate of change of the lifetime with respect to the particle fluence [35,65,139,140]:

$$1/\tau = (1/\tau_0) + K_\tau \phi \tag{108}$$

where K_τ is given by

$$K_\tau = \sigma v_{th} F_e P_R \tag{109}$$

and τ_0 is the initial lifetime, ϕ is the incident fluence (the total number of particles/cm^2, i.e., the flux integrated over time), σ is the capture cross section, v_{th} the thermal velocity, F_e the Fermi probability that the generated recombination center is occupied by a majority carrier, and P_R is the number of centers per centimeter produced by each particle.

It has become more customary to describe the damage coefficient in terms of the minority carrier diffusion length. The relationship (108) then becomes

$$1/L^2 = (1/L_0^2) + K_L \phi, \tag{110}$$

and the rate of change equation is

$$d(1/L^2)/d\phi = K_L. \tag{111}$$

The smaller the value of K_L, the less the diffusion length will decrease with incident particle radiation, and the less the solar cell will degrade.

The second useful measuring parameter is the critical fluence ϕ_c, the number of particles/cm^2 of a given type required to decrease the efficiency (or, the maximum power output) to 75% of its initial value. It might be expected that protons, neutrons, and alpha particles with their heavy masses would have lower critical fluences (and higher K_L's) than electrons of the same incident energy, and such is indeed found to be true; protons, for example, have critical fluences a factor of 10^3–10^4 lower than electrons of comparable energy.

Measured values of K_L and ϕ_c for conventional N/P and P/N Si solar cells, boron and phosphorus doped, are given in Tables 13 and 14. A study of the trends of these measured values and

TABLE 13
Damage Coefficients and Critical Fluences
for Si, Electron Irradiation

Energy	Type	Resist (ohm-cm)	K_L (particle^{-1})	ϕ_C (particles/cm^2)	Ref
300 keV	N/P	1	6.3×10^{-12}	7×10^{16}	141
250 keV	P/N	1	1.5×10^{-10}	3×10^{15}	141
1 MeV	N/P	10	1.7×10^{-10}	8×10^{14}	142
1 MeV	P/N	10	2.6×10^{-9}	4×10^{13}	142
1 MeV	N/P	5	--	7×10^{14}	143
1 MeV	N/P	1	--	1.5×10^{14}	143
1 MeV	N/P	9	8×10^{-11}	7.5×10^{14}	144
1 MeV	N/P	10	5.8×10^{-11}	--	145
1 MeV	N/P	1	1.8×10^{-10}	--	145
1 MeV	P/N	1	2.6×10^{-9}	--	145
1 MeV	N/P	40	2.5×10^{-10}	--	146
1 MeV	N/P	10	4.9×10^{-10}	--	146
1 MeV	N/P	0.5	1×10^{-9}	--	146

TABLE 14
Damage Coefficients and Critical Fluences
for Si, Proton Irradiation

Energy	Type	Resist (ohm-cm)	K_L (particle^{-1})	ϕ_C (particles/cm^2)	Ref
12 keV	N/P	10	--	3×10^{14}	147
32 keV	N/P	10	--	8.5×10^{13}	147
55 keV	N/P	10	--	10^{13}	147
450 keV	N/P	10	--	5×10^{9}	148
2 MeV	N/P	9	8×10^{-7}	9.5×10^{10}	144
7 MeV	N/P	9	4×10^{-7}	3×10^{11}	144
10 MeV	N/P	9	3.5×10^{-7}	4.5×10^{11}	144
30 MeV	N/P	9	3×10^{-7}	7×10^{11}	144
70 MeV	N/P	9	1.8×10^{-7}	9×10^{11}	144
100 MeV	N/P	9	1.6×10^{-7}	1.2×10^{12}	144
155 MeV	N/P	9	1.3×10^{-7}	1.7×10^{12}	144
8.3 MeV	N/P	1	3×10^{-6}	8.5×10^{10}	141
8.3 MeV	P/N	1	7×10^{-6}	5×10^{10}	141
19 MeV	N/P	1	7×10^{-7}	$1-6\times10^{10}$	141
19 MeV	P/N	1	2×10^{-5}	$5-50\times10^{9}$	141

a study of the cited references reveals several interesting
generalizations for conventional solar cells.

(1) Cells of the N/P variety are considerably more radiation tolerant than the P/N variety [141-143,145,146]. This
is true for both electron and proton irradiation, although the
difference becomes smaller for high energy protons.

(2) The radiation tolerance becomes smaller as the base
resistivity is reduced [34,46,141,143,146]. This is unexpected,
since lower resistivities generally mean shorter lifetimes,
and from (108), shorter lifetime devices should be less affected by radiation. This trend is also unfortunate, because lower
resistivities lead to higher predicted starting efficiencies.
It should be kept in mind, however, that the low resistivity-
low tolerance trend has only been well established for boron-
or phosphorus-doped bases. Impurities such as boron readily
associate with radiation-induced defects to form recombination
centers, and the more boron available, the more readily these
centers are produced. Aluminum doping substituted for boron
appears to result in greater tolerance [149-151], presumably
because Al does not behave in the same manner. Madelkorn *et
al.* [149] found that 10 ohm-cm Al-doped cells preserved their
base diffusion lengths under 10 MeV protons considerably better
than boron-doped cells of the same resistivity; they also found
that Cl present in the base of boron-doped cells improved the
radiation tolerance to 1 MeV electrons compared to boron-doped
devices without the Cl, even though the tolerance to proton
bombardment was unaffected by the Cl. The role of impurities
in determining radiation tolerance is a subject which deserves
more attention in the future.

(3) Different particles and particle energies result in
different degrees of damage and degradation. For electrons
with energies less than 1 MeV, ϕ_c and K_L are strong functions
of the energy, as shown for ϕ_c in Fig. 89, although a saturation of ϕ_c with increasing energy above 1 MeV is suggested in
Fig. 89. For protons, ϕ_c and K_L are weak functions of energy.
Moderate energy protons can degrade cells more than high energy
protons; ϕ_c decreases in the energy range of 10-400 keV, reaches
a minimum (highest rate of degradation) at 400-500 keV, and
then increases with increasing energy. ϕ_c and K_L remain relatively constant [144] for protons from 7 to 40 MeV.

Protons have damage coefficients several orders-of-magnitude higher and critical fluences several orders-of-magnitude
lower than electrons; one proton at 1 MeV creates as much
damage as 10,000 1 MeV electrons [144].

Low energy electrons and protons do not penetrate far
into the cell. The physical damage they create therefore is

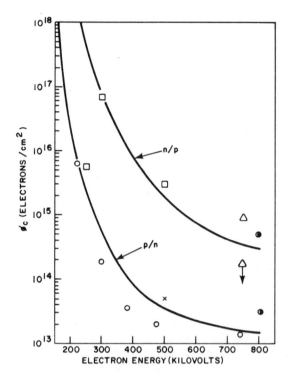

FIG. 89.   Critical fluences ϕ_C versus particle energy for
electron irradiation, Si solar cells.   (After Baicker and
Faughnan [141]; courtesy of the American Institute of Physics.)

close to the surface, and as a result the short wavelength,
high photon energy spectral response is degraded, as shown in
Fig. 90a. High energy particles, on the other hand, penetrate
deeply into the cell and degrade the lifetime and diffusion
length more or less uniformly throughout the base. The long
wavelength, low photon energy spectral response is strongly
reduced by these high energy, penetrating particles, as shown
in Fig. 90b.

(4) There can be a considerable difference in the radiation tolerance between float-zone (FZ) Si and crucible-grown
(CG) (Czochralski) Si. At one time, FZ cells were believed
to be superior to CG cells in radiation tolerance because of
a lower oxygen content. More recently [20,151], the relative
radiation tolerances of the two types of devices are believed
to be more associated with dislocations and with the boron
content, and to a lesser degree with the oxygen content. Float-
zone cells, which normally have a higher starting lifetime

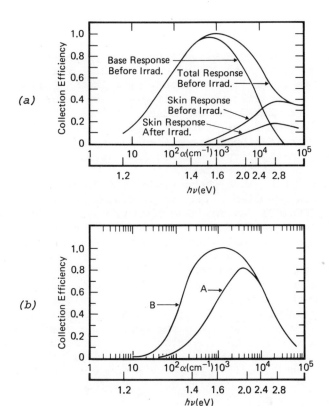

FIG. 90. Relative internal spectral responses of Si cells
before and after particle bombardment: (a) decrease in response
of the top region after low energy (250 keV) electron irradiation; (b) decrease in base response after high energy (8.3 MeV)
proton irradiation (B = before irradiation; A = after irradiation). (After Baicker and Faughnan [141]; courtesy of the
American Institute of Physics.)

than CG devices, are slightly more radiation tolerant than
CG cells, but can undergo significant *further degradation* [151]
when exposed to *AM0 sunlight* after being electron irradiated.
Crucible-grown cells, on the other hand, were found to *improve*
slightly when exposed to sunlight after being electron irradiated. Crabb [151] found that FZ Si cells produced with special
low dislocation density material did not exhibit photon degradation, and neither did normal FZ cells with Al doping instead
of boron. He attributed the photon degradation effect to the
activation of boron-vacancy point defects normally pinned at

dislocation sites. Fischer and Pschunder [20] also investigated photon degradation in FZ and CG solar cells and found that photon-induced degradation often took place even in non-irradiated material, both FZ and CG, and the largest effect was observed in 1 ohm-cm CG devices. The photon-induced degradation could be reversed by annealing at temperatures above 200°C. They attributed the degradation to the destruction of oxygen-impurity complexes; this destruction results in increased numbers of recombination centers. Annealing (below 600°C) generates oxygen-impurity complexes and therefore lowers the number of recombination centers.

It would appear that photon-induced degradation can be a problem in both irradiated and nonirradiated FZ and CG solar cells, but if the FZ devices have low dislocation densities and low oxygen content, they should be 2 to 3 times more radiation tolerant than CG cells and not suffer from photon-induced degradation. Aluminum substituted for boron in the base would also improve the radiation tolerance.

(5) High energy particles affect the short circuit current more strongly than they do the open circuit voltage, while low energy particles (10-400 keV) do just the opposite [148]. High energy particles penetrate into the device and reduce the diffusion length in the base, lowering the photocurrent. Low energy particles create damage closer to the junction edge, reducing the open circuit voltage by a combination of lower shunt resistances and a greater defect density within and near the depletion region.

The effective doping density in the base can be changed due to radiation bombardment. The lattice defects created by the incident particles can act as impurity compensators, removing some of the majority carriers from the appropriate energy band. This process is known as "carrier removal," and it can raise the base resistivity, lower the open circuit voltage, and lower the short circuit current all at the same time.

The "violet cell" with its very thin, relatively lightly doped diffused region is reported to have improved radiation tolerance compared to conventional N/P cells [4,53,152-154]. The critical fluence for 1 MeV electrons for the early violet cells [4] was around $3\times10^{15}/cm^2$, a factor of 3 to 4 higher than conventional 10 ohm-cm cells. Later reports show the violet cell to be superior for proton radiation [152] and neutron irradiation [153] as well.

B.   Lithium-Doped Cells

One of the most significant discoveries of the late 1960's as far as solar cells are concerned was that lithium incorporated in the base of a P/N Si solar cell makes the cell significantly more radiation tolerant than conventional N/P or P/N cells. In the early experiments, Li-doped P/N cells and conventional boron-doped N/P cells were irradiated simultaneously with 1 MeV electrons [3] and 16.8 MeV protons [155]; it was found that both cells degraded initially at about the same rate, but the Li-doped cells "recovered" nearly to their original (preirradiated) properties after bombardment was stopped and the cells were stored for several weeks at room temperature; the cells without Li did not exhibit such recovery behavior. It was conjectured that Li diffuses to and combines with radiation-induced point defects such as vacancies and vacancy-phosphorus complexes. Instead of forming a recombination center as boron does, the Li supposedly neutralized the defect in some way and prevented a degradation in the lifetime.

Today, it is believed that oxygen plays a major role in determining the properties of Li-doped cells [82,83,156], and the concentration of Li relative to the concentration of oxygen largely determines the radiation tolerance and recovery properties. Lithium-doped cells made from oxygen-lean Si (such as FZ material) recover at a fast rate even at room temperature, but tend to be unstable (exhibit fluctuating behavior after recovery), and wide variations in electrical behavior are often observed for units made under identical conditions. Cells made from oxygen-rich Si (such as Czochralski) on the other hand are several orders-of-magnitude slower in recovery than oxygen-lean cells at the same temperature, but are quite stable once recovery has occurred and are uniform in their electrical properties. For environments in which the temperature exceeds about 50°C, oxygen-rich Li cells are superior to both oxygen-lean cells and conventional 10 ohm-cm N/P cells in initial efficiency and in retaining their efficiency after irradiation [157]. For environments from 20 to 50°C, oxygen-rich Li cells recover too slowly to be of much advantage over conventional N/P cells or violet cells, but oxygen-lean devices still provide higher output because of their fast recovery rate. For environments below 20°C, Li doping does not provide any advantage in radiation tolerance over conventional devices [157].

Further study showed that the advantages at temperatures above 20°C of Li-doped cells over boron-doped 10 ohm-cm N/P devices depended on the energy and type of the incident particle. The advantage of Li cells was marginal for low energy electrons (<1 MeV) and protons (<100 keV) which create most

TABLE 15
Critical Fluences of Silicon (Li-doped) P/N
Solar Cells for Proton Irradiation[a]

Energy (MeV)	ϕ_C (before anneal) (particle/cm^2)	ϕ_C (after anneal) (particle/cm^2)
11	7.5×10^{10}	3.3×10^{12}
20	7×10^{10}	5×10^{12}
27	8×10^{10}	5×10^{12}
37	8×10^{10}	3.5×10^{12}

[a]After Anspaugh and Carter [159].

of their damage near the surface, away from the regions of Li
doping in the base.   Berman [82], in summarizing work done on
contract for the Jet Propulsion Laboratory, states that the
best Li-doped cells were only 15% better in efficiency than
10 ohm-cm N/P cells after six months of 1 MeV electron bombardment at low flux rates.   On the other hand, Li cells degraded
at only 1/10 the rate (after recovery) of conventional cells
for 28 MeV electron bombardment, which creates most of its
damage deep in the base.   Similar dramatic advantages were
observed with high energy (>1 MeV) proton and neutron irradiation [82,83,153,158,159].   Table 15 lists the critical fluences
for Li cells irradiated with protons of various energies, as
measured by Anspaugh and Carter [159].   The damage to Li cells
immediately after irradiation, before any recovery takes place,
is equal to or even slightly greater than the damage to conventional N/P cells.   After recovery takes place (annealing
at 60°C for times from 121 hr to 34 days), the Li cells are
considerably better than conventional cells (compare Tables 14
and 15).   In space, where irradiation proceeds constantly and
at a slow rate, recovery goes on simultaneously with irradiation, and the radiation tolerance of Li cells is higher at all
times.
      Both the Li concentration in the base of the cells and
the gradient in Li concentration near the junction must be
closely controlled [83] in order to obtain consistency and
reliability in the recovery and stability of the devices.
Since the gradient in Li concentration near the junction edge
is an easily measured quantity (through the reverse biased
capacitance-voltage relationship), it has become customary to
specify the radiation tolerance, recovery properties, and
stability as a function of this gradient.   (The gradient in

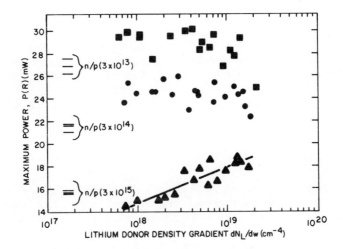

FIG. 91. *Maximum power output after recovery as a function*
of Li gradient, for several fluences of 1 MeV electrons. ϕ:
■ $3\times10^{13}e/cm^2$; ● $3\times10^{14}e/cm^2$; ▲ $3\times10^{15}e/cm^2$. *(After Faith*
[158]; courtesy of the IEEE.)

doping density establishes an electric drift field in the base
near the junction that helps in obtaining high photocurrents
and low dark currents in Li-doped devices [160].) For the best
devices, the Li gradient lies between 5×10^{18} and 4×10^{19} cm^{-4},
with Li concentrations over most of the base of 5×10^{14}-1×10^{16}
cm^{-3}. If the concentration is below this range, there is insufficient Li present to neutralize all the damage created at
high fluences, and recovery from the degradation ceases when
the defect density introduced by the damage becomes comparable
to the Li concentration (carrier removal). When the Li concentration is above this range, the starting efficiency is
lower because of smaller diffusion lengths as in conventional
low resistivity cells (although the unit can withstand higher
doses and can recover at a faster rate). Figure 91 shows the
maximum power output (AM0) after recovery as a function of Li
gradient for increasing fluences of 1 MeV electrons. At low
fluences, units with lower Li gradients (and lower Li densities) are slightly better, but as the dose is increased, devices with high Li content are better. (Figure 91 also shows
the output from conventional N/P cells irradiated at the same
time, and demonstrates the 15-20% improvement obtained with
Li cells for 1 MeV electrons.)
    The initial AM0 efficiencies of optimized Li-doped cells
have been as high as 12.8%, with average values of 11.9%,

compared to 11.5% average values for conventional 10 ohm-cm
N/P cells. The high efficiencies, combined with increased
radiation tolerance for heavy and energetic particles, make
Li P/N cells seem very attractive for space environments compared to conventional N/P cells. The violet cell is higher
in initial efficiency than Li cells, but Li cells continue to
be superior in radiation tolerance at high particle energies
and fluences.

C.   Annealing

The recovery of Li cells when annealed at temperatures
above 60°C can be very dramatic. Conventional N/P devices and
violet cells can also recover upon annealing [46,153], but the
recovery is slight. The recovery is caused by atomic movement
within the crystal with partial elimination of the radiation-
induced lattice disorder and a reduction in the number of recombination centers. Faraday et al. [46] have done extensive
work on the annealing of 4.6 MeV proton damage to N/P boron-
doped cells and found that annealing took place in two distinct
stages; partial recovery occurred when the temperature reached
50-150°C, and subsequent recovery could be obtained only if
the temperature was increased to around 400°C. They also found
that nearly total recovery occurred for 10 ohm-cm cells irradiated with fluxes as high as 10^{12} protons/cm^2 for annealing
temperatures in excess of 500°C, while 1.5 ohm-cm cells, in
addition to their initially greater degradation due to the
higher boron concentration, did not recover nearly as well as
the 10 ohm-cm cells. Fang and Liu [161] found that 1 MeV electron damage at doses as high as 10^{16}/cm^2 could be almost totally
removed by annealing at 400-450°C in about 20 min; the annealing
of electron damage proceeded in only one stage however, in contrast to the two stage proton damage annealing.
The value of annealing at temperatures of 400°C and above
is questionable as far as space applications are concerned,
since it would be difficult and costly to build a system for
raising and controlling a solar cell array to such temperatures
for the necessary time periods, not to mention the incompata-
bility of these temperatures with the present lead-tin solder
metallurgy. There is a need for experimentation in radiation
damage annealing at around 100°C for very long periods of time,
conditions generally met in the space environment at the earth's
orbit.

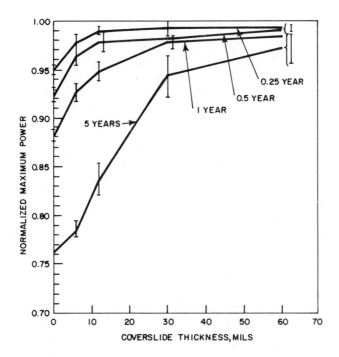

FIG. 92. Degradation of (10 ohm-cm, 2×2 cm, 12 mil thick, boron-doped, N/P Centralab) Si solar cells as a function of time and coverslip thickness for simulated orbital electron irradiation. Error bars represent 95% confidence limits. (After Goldhammer and Anspaugh [162]; courtesy of the IEEE.)

D. Coverslips

Solar cells placed in a space environment would quickly become unusable without some form of protection; unprotected cells would begin to degrade in a matter of days from proton irradiation and in a few months as a result of electron irradiation. The most common means of protecting the cells is to cover them with a transparent sheet of glass, quartz, sapphire, and the like, with a thickness of 1-50 mil, bonded to the cell surface with a transparent adhesive. An alternative is to deposit a layer of SiO_2, Al_2O_3, etc., directly onto the surface ("integral coverslips"), but the thickness of such layers is much less. The higher the density and thickness of a coverslip, the greater is its ability to prevent particles from reaching (and degrading) the cell. Figure 92 illustrates the degradation of 10 ohm-cm N/P Si cells, under electron irradiation simulating the electron flux in synchronous orbit, as a function of coverslip thickness using fused quartz. Thin coverslips can protect solar cells for a short period of time, but thicknesses of around 50 mil are needed to prevent serious loss of power over a five-year period. A similar set of curves would result for proton irradiation, but the curves would be shifted downward in the same time frame because of the larger damage coefficients for protons.

Coverslips go a long way toward minimizing radiation damage problems, but they are not without problems of their own. In addition to the weight they add to the solar cell array, the coverslips themselves can degrade in the space environment. The adhesive has a tendency to darken under ultraviolet light, requiring a UV rejection filter on the coverslip which lowers the short wavelength response of the overall cell. The coverslip itself can become less transparent under particle bombardment due to the formation of color centers [163]. Sapphire and quartz resist this darkening better than glass, but are considerably more expensive. Another material, FEP teflon, is relatively cheap and can be used as both an adhesive and as a coverslip [164,165], but little is known as yet about its behavior in the space environment. One advantage of the FEP as an adhesive and coverslip is that the UV rejection filter is no longer needed [165].

Certain types of impurities added to a coverslip material can improve its ability to resist darkening. Hydrogen added to 12 mil thick glass covers [163] increased their resistance to electron, proton, and neutron radiation in the 1-2.5 MeV range, while cerium added to microsheet glass covers [166] improved their resistance to 1 MeV electrons. The Ce-doped coverslips had the same advantage as FEP teflon in cutting off UV light below 3600 Å, while conventional (and expensive) multilayer rejection filters begin to cut off light at around 4000 Å, so that the Ce covers allow more light to reach the cell while still preventing UV-induced degradation of the adhesive.

Integral covering layers [167-169] deposited directly on the surface by sputtering, thermal evaporation, or vapor growth are less costly than adhesive mounted covers but generally have to be made thinner due to stresses on the cell from thermal expansion and bonding differences between the two materials. Fused silica coatings are usually kept 2 mil or less in thickness; thicker coatings cause bending and increase the risk of breaking the cell during handling. Certain types of glass, such as Corning 7070, are better matched to the properties of Si [167,168], and have been deposited to a thickness of up to 12 mil. The 7070 glass has reportedly excellent optical transmission properties and excellent resistance to darkening [168], and is low in cost. Other types of glasses integrally bonded to Si solar cells have been discussed by Rauch *et al.* [170].

E. Drift Fields

It has been suggested that electric drift fields incorporated in the base of Si solar cells might improve their radiation tolerance by adding a drift component to the flow of carriers, making them somewhat less dependent on the base diffusion length [34,37,40,41]. However, it is somewhat ambiguous whether this is true or not. First of all, there is a trade-off between the high doping level ratios from the junction edge to the back of the cell needed to obtain a substantial electric field and the decrease in mobility, lifetime, and diffusion length which accompanies the increasing carrier concentration. Second, the damage coefficient increases and critical fluence decreases with the increasing boron concentration [171]. Experimentally, the devices that have been made, with 10 to 60 μm wide drift field regions, have been somewhat poorer in initial efficiency and very slightly improved in radiation tolerance to 1 MeV electrons [41,171] compared to conventional N/P devices.

The situation for a BSF device is clearer. Here the doping is usually uniform over most of the base and high doping levels exist only at the back. There is no drift component in the base to aid the flow of carriers, but some improvement in radiation tolerance is predicted anyway because of the carrier confining principle (a BSF device is somewhat less dependent on base diffusion length than a conventional cell). As long as the base thickness is less than a diffusion length, higher open circuit voltages and short circuit currents will be obtained than in a conventional cell of the same thickness; therefore, the change in efficiency with increasing fluences of penetrating particles will be less than in conventional cells with Ohmic back contacts. Once the base diffusion length has been degraded to less than the base thickness, a BSF cell will degrade at the same rate as a normal N/P cell.

F. Other Solar Cells

Solar cells made from direct bandgap materials are usually more radiant tolerant than devices made from indirect gap materials, particularly for penetrating particles. The higher tolerance arises from the high absorption coefficient and short

TABLE 16

Critical Fluences for GaAs and Si Solar Cells[a]

Particle, Energy (MeV)	ϕ_C (part/cm^2) GaAs, P/N	ϕ_C (part/cm^2) Si, N/P
0.8,electrons	1.1×10^{15}	1.3×10^{15}
5.6,electrons	2.7×10^{14}	3.0×10^{13}
0.1,protons	$\sim 1 \times 10^{12}$	$\sim 1 \times 10^{13}$
0.4,protons	$< 10^{11}$	1×10^{11}
1.8,protons	2.4×10^{12}	$\sim 1 \times 10^{11}$
17.6,protons	5.7×10^{12}	4×10^{11}
95.5,protons	$> 2 \times 10^{12}$	7×10^{11}

[a]After Wysocki [172].

lifetimes usually found in direct gap materials. Damage created more than a few times α^{-1} beneath the surface has no effect on photocurrent collection, since there are no carriers generated deeper than this. Also, damage created more than a few diffusion lengths from the junction edge has no effect on either the photocurrent or dark current. For GaAs solar cells, for example, any lattice damage created more than 6 μm or so below the surface has no effect on the power output.

On the other hand, low energy particles with ranges of only several microns can degrade direct bandgap solar cells as fast or faster than other types of cells. Table 16 shows the critical fluences of GaAs solar cells for electron and proton irradiation, together with fluences for Si N/P devices measured under the same conditions. The GaAs cells are more than an order-of-magnitude more tolerant to high energy electrons and protons, but are very susceptible to low energy protons because of the heavy damage created in the first micron below the surface. A thin coverglass will screen out low energy particles and should make GaAs cells more tolerant than Si devices at all particle energies.

Both single crystal and thin film CdS solar cells are highly radiation tolerant, although their starting efficiencies are much lower than Si or GaAs devices. The absorption coefficient of CdS is so high [101] that light passing through the Cu_2S and having energy greater than 2.4 eV is absorbed within a micron or so of the CdS-Cu_2S interface; most of the damage created by penetrating, high energy particles is therefore of no importance. The damage to the Cu_2S from high energy particles is minimal because of its small thickness. Low energy particles have slightly more effect, since their range of 1000 Å to 1 μm covers the active regions of the Cu_2S-CdS device.

Van Aerschodt *et al.* [173] have subjected thin film CdS
devices to 100-700 keV proton irradiation, and found critical
fluences greater than 10^{15} cm^{-2} for 100 keV particles, more
than two orders-of-magnitude higher than in Si or GaAs solar
cells under comparable conditions.

G. Summary

When solar cells are irradiated with energetic electrons,
protons, and other particles present in the space environment,
a degradation in performance can occur as a result of the damage
to the lattice. Low energy particles create damage close to
the junction, and therefore raise the dark current and lower
the open circuit voltage. High energy particles penetrate far
into the base and lower the base lifetime, decreasing the short
circuit current. Boron-doped Si cells with higher base resistivities have higher radiation tolerance than lower resistivity
cells. Devices made from low dislocation density FZ material
are more tolerant than devices made from CG material.

Lithium incorporated into the base of P/N Si cells greatly
improves their radiation tolerance. If conventional N/P cells
and Li-doped P/N cells are irradiated simultaneously, both degrade at about the same rate but the Li cells "recover" to
nearly their initial values. Oxygen-lean Li-doped cells recover at a fast rate but are then unstable in behavior; oxygen-
rich Li-doped cells recover slowly but are stable. The recovery rates are also a function of temperature. For temperatures
above 50°C, the recovery in both oxygen-rich and oxygen-lean
cells is fast enough to be useful; below 20°C, the recovery in
both cells is too slow to be useful.

Drift fields in the base of Si cells may improve their
radiation tolerance somewhat due to the aiding drift force on
minority carriers. A BSF improves the radiation tolerance of
thin cells by making them slightly less dependent on the base
diffusion length than conventional devices.

To prevent any solar cell from degrading rapidly in the
space environment, it is necessary to place a coverslip over
the front surface to minimize the number of particles reaching
the device. Many types of coverslips and the adhesive used to
bond them to the cell will darken under UV radiation, requiring
a UV rejection filter to prevent this problem. Cerium-doped
microsheet and FEP teflon coverslips resist UV darkening, however, giving them a considerable advantage over other types.
Integral coverslips deposited directly onto the surface of the
solar cell normally have to be kept thin (<2 mil) to prevent
stresses on the cell, but certain glasses can be deposited up
to 12 mil thick without harmful effects.

CHAPTER 8

Temperature and Intensity

Most solar cells made in the past were designed for near-earth operation, implying an input power level of around 135 mW/cm^2 and a working temperature of 50-60°C. There are also important space applications for cells operating under greatly different conditions, ranging from large solar distances such as Jupiter's orbit, where the ambient temperature is -120 to -130°C and the input intensity is 5 mW/cm^2, to short distances such as the Venus or Mercury orbits, where the temperature exceeds 140°C and the intensity is 250 mW/cm^2 or greater. To optimize solar cell operation at each extreme, it is necessary to understand the behavior of solar cells as a function of temperature and intensity.

Temperature and intensity considerations are also important when considering the use of solar cells for large-scale economical power generation on earth. Sunlight can be concentrated by a factor of 100 or more at a cost below the cost per unit area of most solar cells, and the use of sunlight concentration will probably make solar energy conversion via solar cells an attractive alternative to other available means of generating power by the 1980's.

A. Variable Temperature, Constant Intensity

1. MATERIAL PARAMETERS

The important material parameters which determine the behavior of solar cells as a function of temperature are the intrinsic carrier density n_i, the diffusion length (through the lifetime and mobility), and the absorption coefficient. The intrinsic carrier density plays a large role in determining the open circuit voltage; V_{oc} *decreases* with temperature mostly because n_i, and consequently the dark current, increases strongly with increasing temperature. The decrease in V_{oc} is

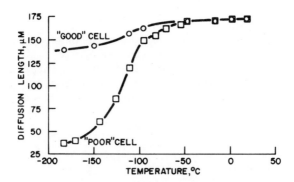

FIG. 93. Variation of diffusion length with temperature for two categories of Si solar cells. (After Mandelkorn et al. [174]; courtesy of the IEEE.)

partly offset by an *increase* in the short circuit current with increasing temperature, due to the improved lifetimes at higher temperatures in materials dominated by ionized impurity scattering and due to a shift in the absorption edge (bandgap) to lower energies.

The minority carrier mobility in the base of a solar cell is determined by a parallel combination of lattice scattering (acoustical phonon scattering), which varies at $T^{-3/2}$, and ionized impurity scattering, which varies as $N^{-1} T^{+3/2}$. For Si cells with doping levels of around 10^{17} cm^{-3} or less, the mobility in the base decreases somewhat with increasing temperature, while the diffusion coefficient, which is $(kT/q)\mu$, is nearly constant with temperature. For GaAs cells, optical phonon scattering is important ($\propto T^{+1/2}$) and the net mobility in the base is nearly constant with temperature, while the diffusion coefficient increases as T^{+1}.

The lifetime is a function of temperature through the thermal velocity (which varies as $T^{+1/2}$) and the capture cross section (which can have either a positive or negative temperature coefficient depending on the nature of the recombination center). In Si, the lifetime generally increases several fold with increasing temperature in the -150 to +150°C range for both n- and p-type samples [18], while in GaAs, the lifetime increases more strongly [28] (rising from 2×10^{-10} sec at 100°K to 1.5×10^{-9} sec at 300°K for a 10^{18} cm^{-3} Zn-doped wafer, for example).

These changes in mobility and lifetime cause the diffusion length in Si and GaAs solar cells to improve with increasing temperature, as demonstrated by Mandelkorn *et al.* [174] for

10 ohm-cm Si cells in Fig. 93. The improvement in diffusion
length in the temperature range of -200 to +25°C amounts to
around 15% for good Si devices, but the change in diffusion
length is much stronger for cells classified as "poor"; this
strong variation in L for poor cells has been attributed to
abnormally high dislocation densities in these devices [174].
The improvement in diffusion length with increasing temperature
for good GaAs cells is larger than for Si cells because of the
larger change in lifetime in GaAs.

2. DEVICE PARAMETERS

    The changes in short circuit current with temperature
for 1 ohm-cm P/N and 2 ohm-cm N/P Si cells at several incident
intensities are shown in Fig. 94, as presented by Yasui and
Schmidt [52]. The photocurrent increases slightly with increasing temperature, partly due to the improvement in base
diffusion length and partly due to the shift of the absorption
edge to lower energies, both of which improve the long wavelength spectral response. The improvement in photocurrent
with increasing temperature is even stronger for GaAs cells
(see Fig. 98), probably due mostly to the shift in the absorption edge [175] but partly to the increasing diffusion length
also.
    The open circuit voltage decreases with increasing tem-
perature in a more-or-less linear fashion for Si and GaAs
solar cells, as shown in Figs. 95 and 98. This decrease in
the open circuit voltage is due to the strongly increasing
dark current. The dark current is composed of the injected
current J_{inj}, the depletion region recombination current J_{rg},
and in some cases a tunneling current J_{tun}. The injected
current varies as $n_i^2 \propto \exp(-qE_g/kT)$, while J_{rg} varies as
$n_i \propto \exp(-qE_g/2kT)$; both of these currents increase strongly
with increasing temperature. Tunneling currents are largely
independent of temperature. Tunneling is not important in
Si cells with base resistivities above 0.1 ohm-cm or in P/N
GaAs cells with resistivities above 0.01 ohm-cm, but tunneling
usually dominates heterojunction devices such as the Cu_2S-CdS
cell and causes the open circuit voltage to remain fairly
constant with temperature in these cells.
    Measured open circuit voltages decrease with temperature
at a rate of about 2.5 mV/°C for 10 ohm-cm Si cells and 2.2
mV/°C for 2 ohm-cm Si devices [52,176], while for GaAs solar
cells the rates are generally between 1.9 and 2.2 mV/°C [6,9].
    The FF decreases with increasing temperature above 200°K,
as shown in Figs. 96 and 98. Part of this decrease is due to

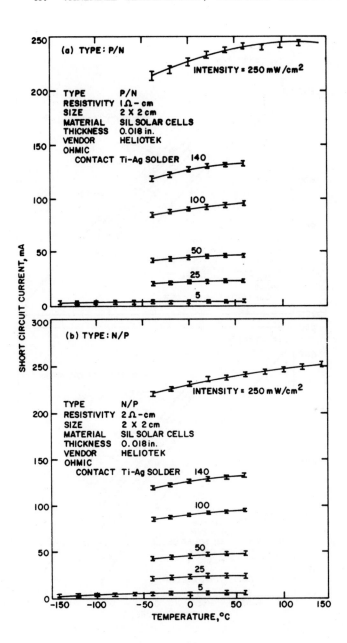

FIG. 94. Short circuit current of 2×2 cm Si solar cells as a function of temperature and intensity. (After Yasui and Schmidt [52]; courtesy of the IEEE.)

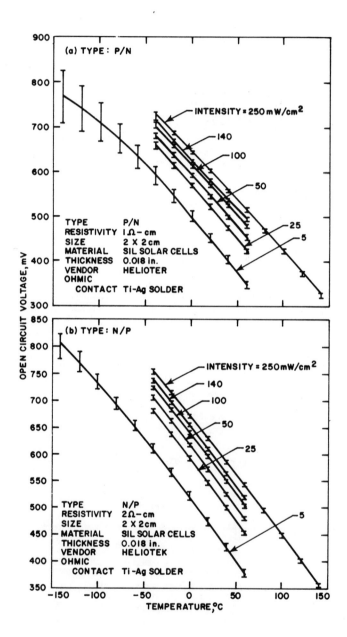

FIG. 95.   Open circuit voltage of Si solar cells as a function of temperature and intensity.   (After Yasui and Schmidt [52]; courtesy of the IEEE.)

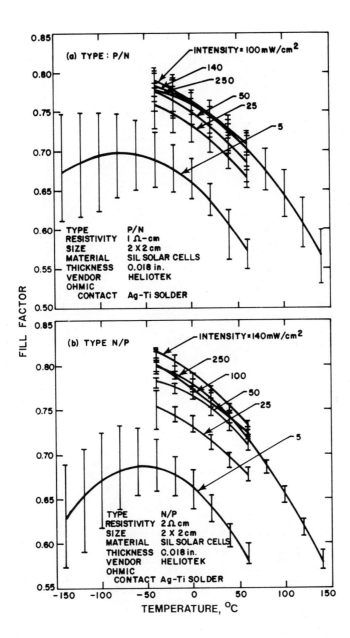

FIG. 96. Fill factor of Si solar cells as a function of
temperature and intensity. (After Yasui and Schmidt [52];
courtesy of the IEEE.)

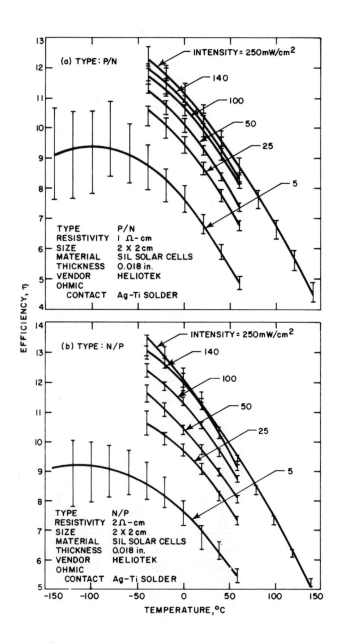

FIG. 97. Efficiency of Si solar cells as a function of
temperature and intensity. (After Yasui and Schmidt [52];
courtesy of the IEEE.)

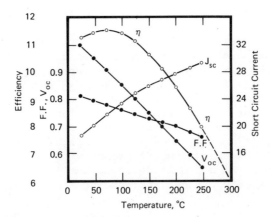

FIG. 98. *Measured parameters of Ga$_{1-x}$Al$_x$As-covered GaAs solar cells as a function of temperature. (After Hovel and Woodall [177].)*

the lower open circuit voltages and part to the increasing "softness" (roundness) in the knee of the I-V curve as temperature increases in the exp(qV/AkT) term. The saturation and slight decrease in FF for temperatures below 200°K might be caused by an increasing bulk resistance or contact resistance at low temperatures, although other factors may also be involved.

As a result of the decrease of V_{OC} and FF with increasing temperature, partly offset by the improvement in I_{SC}, the efficiency normally decreases with increasing temperature, as shown in Fig. 97 for Si solar cells. The rate of change near room temperature ($\Delta\eta/\Delta T$) is around 0.04-0.06%/°C for conventional Si cells with 1-10 ohm-cm resistivities [6,140]. Violet cells and thin BSF cells have higher efficiencies than conventional cells at a given temperature, but the rate of change should be about the same.

The small decrease in efficiency of GaAs solar cells with increasing temperature is one of the main advantages of these cells. In some cases, the efficiency actually increases slightly with increasing temperature, as shown for the Ga$_{1-x}$Al$_x$As-coated GaAs p-n junction in Fig. 98 [177]. This is probably due to the larger increase in J_{SC} compared to the decrease in V_{OC} at temperatures up to around 75°C. The AM0 efficiency of this cell extrapolates to 6% (uncorrected for contact area) at 300°C. By contrast, Si solar cells have very low efficiencies above 200°C and are virtually unusable at 300°C. The rate of decrease of efficiency with temperature

($\Delta\eta/\Delta T$) in GaAs cells is 0.02-0.03%/°C [6,9] for temperatures above 100°C.

CdS solar cells have a much smaller temperature dependence below room temperature than other types of devices, mostly because of the dominance of the dark current by tunneling. Gill and Bube [74] have reported that V_{oc} and I_{sc} in single crystal CdS devices were initially constant between 135 and 295°K, but after heat treatment (250°C for 1 min) variations began to appear, with V_{oc} decreasing from 0.6 V at 113°K to 0.48 V at 288°K and I_{sc} undergoing a slight increase in the same range. The change in efficiency with temperature is smaller than in Si and GaAs devices, but Cu_2S-CdS cells cannot be used above 80°C without introducing stability problems (see Chapter 9).

The temperature behavior of Schottky barrier and hetero-junction solar cells should be qualitatively about the same as in p-n junction devices [177a]. The long wavelength spectral response and the photocurrent should improve slightly with in-creasing temperature due to the increasing lifetime in the base. The open circuit voltage in Schottky barriers should decrease strongly with increasing temperature, as predicted by the $T^2 \exp(-q\phi_b/kT)$ term in the dark current expression (Eq. (85)). In heterojunctions whose dark currents are determined mainly by thermal injection, the open circuit voltage decreases with temperature in the same way as in conventional p-n junctions, due to the n_i^2 dependence. In heterojunctions whose dark cur-rents are determined mainly by tunneling, the open circuit voltage and efficiency are slightly less temperature dependent than in conventional p-n junctions, but the absolute values of V_{oc} and η at any temperature are also less than they would be for heterojunctions dominated by thermal injection currents.

B.   Variable Intensity, Constant Temperature

The optimum design of a solar cell for operation at low intensities is quite different from the optimum design for high intensities. If the temperature is held constant, the current output decreases linearly with decreasing intensity while the voltage output remains about the same (decreases logarithmically). The series resistance is relatively unimpor-tant at low intensities, but the shunt resistance has a strong effect, since a leakage current comparable to the photocurrent strongly reduces both the voltage output and the fill factor. As the intensity is increased at constant temperature, the current output increases linearly but the voltage output still remains about the same (increases logarithmically); the shunt

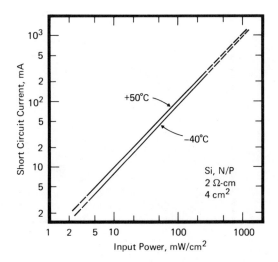

FIG. 99. Short circuit current of Si N/P solar cells as a function of input intensity. (Data from Yasui and Schmidt [52].)

resistance becomes unimportant at high intensities but the series resistance can have a drastic effect. For low intensity operation it is important that the device have high junction perfection; for high intensity operation it is important that the device have an exceptionally low series resistance (many grid lines) and behave well at elevated temperatures.

The short circuit current is proportional to the incident intensity over many orders-of-magnitude, provided that the series and shunt resistances are negligible and that the spectral distribution of the incident light remains the same. Figure 99 shows the photocurrent for the 2 ohm-cm N/P Si device of Fig. 94; the current is linear from 0.04 to 2 suns, the limit of the measurement. Luft [178] has observed a linear increase in photocurrent up to 20 suns intensity, and has discussed the importance of increasing the number of contact grids to minimize the series resistance. Hovel and Woodall [9] observed a linear temperature dependence of I_{sc} from 0.01 to about 1 sun intensity in GaAs solar cells with $Ga_{1-x}Al_xAs$ windows, and Davis and Knight [179] observed linearity up to 1000 suns in similar GaAs cells. At sufficiently high intensities such that $\Delta n = \Delta p = Na$, i.e., when the photogenerated carrier density becomes comparable to the base doping level, the base lifetime may increase and the photocurrent may increase superlinearly with intensity [179a], provided series resistance effects are small.

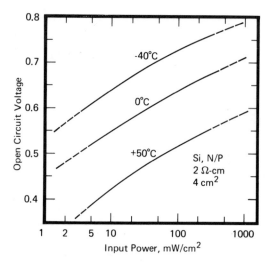

FIG. 100.   Open circuit voltage of Si N/P solar cells as a
function of input intensity.   (Data from Yasui and Schmidt [52].)

The open circuit voltage increases logarithmically with
increasing intensity, as shown in Fig. 100.   This logarithmic
dependence is predicted by Eq. (1) taken together with the
linear dependence of the short circuit current I_{sc} on intensity.
The FF also increases with intensity provided the series resistance is negligible.

As a result of the variations in V_{oc} and FF, the efficiency
increases with increasing intensity, as shown in Fig. 101 (the
linear change in I_{sc} with intensity does not contribute to the
change in efficiency).   In theory, the efficiency would continue
to increase logarithmically with intensity up to the point where
high injection level effects begin to occur, after which the
V_{oc} and η may saturate [179a] or even decrease with further increasing intensity.   In practice the effects of series resistance
usually become dominant before this intensity is reached, and
the FF will begin to decrease with increasing intensity rather
than increase.   The efficiency will then begin to decrease as
the degradation in FF due to series resistance outweighs the
improvement in open circuit voltage with increasing intensity.

C.   General Considerations

If solar cells are used on earth, then the temperature
of the cells and the intensity at which they are operated can

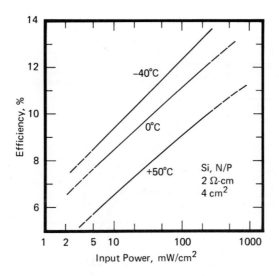

FIG. 101. *Efficiency of Si N/P solar cells as a function of input intensity. (Data from Yasui and Schmidt [52].)*

be semi-independent; one can concentrate sunlight on the cells by a factor of many solar intensities while providing external cooling, for example. In space the temperature and intensity are directly related. At large distances from the sun, the temperature and intensity are both low, while near the sun they are both high. This tends to aggravate the difficulties of operating far from or near the sun; solar cell arrays close to the sun, for example, have reduced efficiencies due to the high temperature, and more severe problems with series resistance due to the higher intensity. The losses due to series resistance can be minimized by increasing the number of fingers in the contact grid (the width of the fingers must be reduced to prevent increasing the contact area loss, however). Present-day solar cells with six contact fingers are designed for operation at about 1 solar intensity (135 mW/cm^2), with net currents of 32-35 mA/cm^2 at the maximum power point and a series resistance R_s of 0.25 ohm for a 4 cm^2 cell. Under these conditions, 30 to 35 mV are dropped across R_s. Magee *et al.* [180] have calculated that a 24-finger structure replacing the common six-finger design would lower the series resistance by more than a factor of 4, which would be adequate for operation at 5 solar intensities or less (675 mW/cm^2). Luft [178] has shown that a 13-grid structure in a 2 cm^2 cell operates satisfactorily at 20 solar intensities (2.7 W/cm^2), while a 5-grid

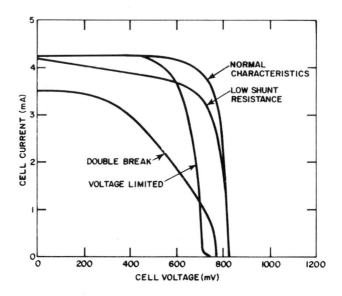

FIG. 102. Examples of anomalous current-voltage behavior in Si solar cells at -130°C and at low light intensities. (After Luft [176]; courtesy of the IEEE.)

device had greatly reduced output because of the high series resistance.

The behavior of Si cells at low temperatures and intensities turns out to be an even greater problem than at high temperatures and intensities. Some of the problems encountered at low intensity/temperature conditions are indicated in Fig. 102. First, a substantial decrease in short circuit current can occur in some devices, apparently caused by a high dislocation density and the effect of dislocations in lowering the diffusion length [174]. Second, the shunt resistance becomes increasingly detrimental [181] as the intensity is reduced and the photocurrent consequently becomes comparable to the diode leakage (shunting) current. Third, the I-V characteristic begins to behave anomalously, an effect described alternately as the "broken knee," "bent knee," or "double slope" phenomenon [181-183]; this phenomenon greatly lowers the FF and reduces both the open circuit voltage and short circuit current to some degree. Payne and Ralph [181] have narrowed the broken knee effect down to a degradation in the performance of the portion of the cell in the vicinity of the stripe contacts, but as yet no explanation of the phenomenon has been well established.

Finally, it appears that the normal contact to the back of the cell, which is reasonably Ohmic at room temperature, is capable of acting like a Schottky barrier at low temperatures [176,181,183], and the relative effect of this barrier becomes worse as the intensity is reduced. This Schottky barrier at the back contact lowers the power output and efficiency by producing a voltage in opposition to the junction photovoltage, resulting in a lower net open circuit voltage. The barrier effect can be prevented [181,183] by adding a BSF or by sintering the back contact at 550-570°C for 20-30 min. The improvement in low temperature device behavior after sintering is impressive [183], and the improvement is even greater for a BSF. It should also be kept in mind that beneficial effects will be gained from an improved back contact at high temperatures and intensities as well as at low temperatures and intensities.

D. Summary

The behavior of solar cells at both high and low temperatures and intensities is important for space applications and for terrestrial applications where sunlight concentration might be involved.

The diffusion lengths in Si and GaAs increase with increasing temperature, and the short circuit current improves with increasing temperature due to this improvement in diffusion length and due to a shift in the absorption edge to lower energies. The increase in photocurrent is higher in GaAs cells than in Si cells. The open circuit voltage in both Si and GaAs cells decreases strongly with increasing temperature due to the large increase in the dark current. The efficiency of Si cells decreases with increasing temperature (due to the lower V_{oc} and FF) at a rate of 0.04-0.06%/°C ($\Delta\eta/\Delta T$), and Si cells become practically unusable at temperatures above 200°C. The efficiency of GaAs cells may remain constant or even improve slightly with temperature up to 80°C, then decreases at around 0.02-0.03%/°C at higher temperatures. GaAs cells are usable at temperatures up to about 350°C. Heterojunction and Schottky barrier cells (except for Cu_2S-CdS) behave in about the same manner with temperature as conventional p-n junction cells. Cu_2S-CdS devices depend very weakly on temperature due to the dominance of the dark current by tunneling.

The short circuit current increases linearly with increasing intensity, while the open circuit voltage increases logarithmically. At low intensities, the shunt resistance and junction perfection become very important, while at high

intensities the series resistance is the most important consideration and the contact grid design becomes highly important.
The efficiencies of Si and GaAs cells increase slightly with
increasing intensity up to the point where the decrease in FF,
due to the series resistance effects, outweighs the improvement
in open circuit voltage.

The operation of solar cells at low temperatures and intensities simultaneously may be difficult due to the appearance
of a Schottky barrier effect at the back contact, due to increased shunt resistance problems, and due to a "broken knee"
effect. The barrier effect can be eliminated by providing a
BSF or by careful contact sintering.

For terrestrial applications, sunlight can be concentrated
by a factor of several hundred on Si solar cells and several
thousand on GaAs cells. Heat rejection then becomes a major
problem. The efficiency will tend to drop due to higher operating temperatures, but this will be partly offset by the beneficial effect of high intensities on V_{oc} and FF.

CHAPTER 9

Solar Cell Technology

The behavior of a solar cell is influenced strongly by the technologies that go into making the cell, and the optimum design for a device depends on the application for which it is intended. For space applications, N on P Si cells incorporating a BSF and a violet cell front might be used, or alternately, a P/N Li-doped device might be used because of its high radiation tolerance. For missions near the sun, GaAs cells with $Ga_{1-x}Al_xAs$ layers have the highest potential because of their excellent high temperature properties and good radiation tolerance. For terrestrial applications, CdS cells, or cells made from ribbon Si or thin film Si or GaAs have the greatest chance of meeting the efficiency per unit cost figure necessary to make solar cells competitive with other means for generating large amounts of electric power.

In this chapter, the technologies that are used to produce solar cells will be briefly discussed, including crystal growth, diffusion, electrical contacting, doping, and the deposition of antireflective coatings.

A. Silicon Solar Cells

1. CRYSTAL GROWTH

The crystal growth methods of most interest for solar cell work are the Czochralski, float-zone (FZ), and ribbon methods and chemical vapor deposition. In the Czochralski method, molten Si is contained in a crucible at a temperature just above the melting point and a temperature gradient is established in the vertical direction. A small seed crystal is introduced into this melt, and then simultaneously rotated and pulled out of the melt, producing a crystalline ingot by freezing at the solid-liquid interface. Close control of both the vertical and horizontal temperature profiles is necessary to minimize defects and prevent doping nonuniformities in the

TABLE 17
Distribution Coefficients of Impurities in Si[a]

Impurity	Coefficient	Impurity	Coefficient
B	0.8	P	0.35
Al	0.002	As	0.30
Ga	0.008	Sb	0.023
In	0.0004		

[a]After Rhodes [184].

crystal. The input power to the melt is usually supplied by
rf induction heating, and the crystals are grown at rates of
10^{-4}-10^{-2} cm/sec.

In this growth method as in others, the dopant impurity
can be added directly to the melt, and some of the dopant will
then be incorporated into the growing solid. The ratio of the
impurity content in the solid to that in the melt under equilibrium conditions is known as the distribution coefficient.
Distribution coefficients for several common dopants in Si are
shown in Table 17. A low distribution coefficient (less than
unity) indicates that the amount of impurity that must be
added to the melt must be larger than the amount desired in
the final crystal, and generally also means that the crystal
will be nonuniformly doped along its length due to impurity
build-up in the melt as the growth proceeds. The relatively
high distribution coefficient of boron is one of the main reasons for its use rather than Al or Ga as the dopant in p-type
Si.

One of the major difficulties associated with the Czochralski technique is that of finding an inert crucible material
to contain the molten Si, one that does not react with the
melt and contaminate it. Carbon has often been used, but the
probability of incorporating carbon into the grown Si is high.
Carbon crucibles are also fairly porous, and this gives rise
to various types of impurities which can seep into the melt.
Currently, the best crucible material appears to be vitreous
silica, as long as care is taken to ensure that the silica is
free of unwanted contaminants. The slight solubility of SiO_2
in molten Si makes it almost inevitable that the grown crystal
will contain a high concentration of oxygen, and this oxygen
content together with other impurities from the crucible cause
the lifetime and mobility in Czochralski Si to be slightly
lower than in FZ Si of the same resistivity [20].

In the float-zoning technique, a zone of molten Si is
slowly passed along the length of a Si ingot held in a vertical

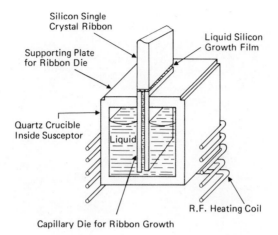

Silicon Single
Crystal Ribbon

Liquid Silicon
Growth Film

Supporting Plate
for Ribbon Die

Quartz Crucible
Inside Susceptor

Liquid

R.F. Heating Coil

Capillary Die for Ribbon Growth

FIG. 103.   *Crucible used for the EFG growth of ribbon Si.*
(After Bates et al. [185]; courtesy of the IEEE.)

position. As the zone progresses, material melts at one
boundary and recrystallizes at the other, with the melt held
in place by surface tension. The regrown Si crystal emerges
with high purity, due both to the absence of a crucible in
contact with the melt and to the purifying action of the molten
zone in preventing impurities with low distribution coefficients
from entering the growing crystal. The room temperature lifetime and mobility tend to be higher than in Czochralski material, and the oxygen content is lower, but the FZ material
generally has a higher dislocation density.

     The higher lifetime leads to larger diffusion lengths,
and consequently leads to devices with higher photocurrents
than devices made from Czochralski Si, and the lower oxygen
content leads to faster recovery from radiation damage in Li-
doped FZ cells compared to Czochralski cells (see Chapter 7).
However, the high dislocation density of FZ material can lead
to both poor lifetimes and loss of photocurrent at low temperatures [174]. The high dislocation densities can also cause
solar cells made from boron-doped FZ material to degrade when
exposed to sunlight [20,151], particularly if the cells have
been irradiated with high energy electrons or protons. (This
photon-induced degradation might be due to the generation of
boron-vacancy point defects at dislocations, although it seems
likely that oxygen also plays a role in determining the density
of recombination centers [20,151].)

     Another technique for growing Si bulk crystals that is
receiving increasing attention is the ribbon growth technique

FIG. 104.   Silicon ribbon.   (Photo by courtesy of A. I. Mlavsky.)

[185-188], by which a thin, ribbon-like crystal is pulled from a Si melt at a very fast rate (in./min). Such ribbons can be grown by the edge-defined method (EFG, which stands for edge-defined, film-fed growth) [185-188], or by the earlier dendritic web technique [189-189c]. The web ribbons differ from the EFG ribbons by the presence of two dendrites bounding the edges of the web Si. Figure 103 shows a cross section of a crucible used in growing EFG ribbons [185]. The molten Si is replenished at the solid-liquid interface by capillary action, and the growing ribbon takes the shape of the die surface. The ribbon is thin enough that wafering (necessary for Czochralski and FZ material) can be eliminated, and the smoothness of the surface can be good enough to eliminate the need for polishing. Many ribbons (20 or more) can be simultaneously grown from the same crucible, and the molten Si charge in the crucible can be continually replenished to

obtain ribbons as long as desired. A photograph of a growing
EFG ribbon is shown in Fig. 104.

Currently, the ribbon technique is probably the least
expensive method for producing Si crystals, and it has a high
potential for making solar energy conversion using solar cells
cost competitive with other methods of generating large amounts
of electrical power [190]. Mlavsky of Tyco Laboratories has
estimated [187] that the cost of the Si ribbon can eventually
be brought down to less than 1¢/in^2, and that the cost of producing electricity using ribbon solar cells could be as low as
$375/kW, equivalent to roughly three to four times the cost of
the ribbon Si itself. Concentration of sunlight could bring
the cost down even further. (By way of comparison, the cost
of generating electricity using fossil fuels is $350-500/kW.)

To realize the potentials of ribbon Si for solar cell use,
wide (1-2 in.), flat, thin (4-6 mil) ribbons must be produced
with suitable minority carrier lifetime and mobility. In the
past, it has been difficult to grow ribbons without twin boundaries and high dislocation densities [185a] which tend to lower
the lifetime. The die material is also a major problem. Graphite and SiC-coated graphite tend to produce carbon and SiC
particles in the ribbon crystal [185a], and deterioration of
a graphite die by reaction with the molten Si can eventually
lead to polycrystallinity [188]. An alternative die material,
which is wet by molten Si but does not react with it, is ac-
tively being sought.

In spite of these problems, excellent device results have
already been achieved with ribbon Si. Crystals 40 cm long by
2 cm wide with 0.3-0.5 mm thickness (12-20 mil) are routinely
grown, at rates of 1-2.5 cm/min [188], and solar cells with
10% efficiency at AM1 have been made from selected areas of
this material [191]. Electron diffusion lengths of 30-150 μm
were measured in the 1-5 ohm-cm p-type ribbon material and
short circuit currents of up to 18.5 mA/cm^2 were obtained in
the finished devices, with open circuit voltages of 0.51-0.53 V.
Ribbon Si has also been used as a substrate for the growth of
epitaxial layers [192], and p-n junctions of reasonable quality
have been made in such epitaxial layers even when polycrystalline ribbon substrates were used.

The growth of Si thin films is another promising technique
for impacting terrestrial energy needs. It has already been
shown in Chapter 5 that 10 μm thick Si films can have AM1
efficiencies above 10% provided that the grain size exceeds
5-10 μm and that the grains are fibrously oriented. The low
cost projections for devices made in such films depend on the
ability to meet these criteria using relatively inexpensive,
plentiful substrates such as Al, steel, glass, or the like.

There has been a great deal of work on polycrystalline
Si layer growth by chemical vapor deposition in recent years
[93,96-99,193-198]. Silicon films can be grown using SiH_4,
$SiCl_4$, $SiHCl_3$, or similar starting materials at temperatures
of 300-1000°C, yielding growth rates of several hundred to
several thousand angstroms per minute. The growth rate and
grain size increase with increasing temperature, and thicker
films tend to yield larger grain sizes than thinner films.
The growth rate can be increased by using N_2 or He instead of
H_2 as the carrier gas [196,197], and by adding diborane to the
gas phase to dope the films with boron [93]. Arsine added for
As doping has the opposite effect of lowering the growth rate
for otherwise equal conditions. Fibrously oriented grains
about 0.2-0.5 μm in size and oriented in the <110> direction
have been obtained on SiO_2-covered Si substrates [93,196].
      Chu [89,194] has described the growth of 10 μm thick Si
films on steel substrates at 800°C and above. Silicon reacts
with iron at such temperatures, necessitating a "diffusion"
barrier of an inert material between the substrate and Si film.
Tungsten and titanium were found to be unsatisfactory; the
best diffusion barriers found so far have been SiO_2 or boro-
silicate. Grain sizes of 5 μm have been obtained on borosili-
cate-steel substrates at 1000°C, and 2 μm grain sizes have
been obtained at 900°C. Somewhat larger grains (10-15 μm) were
obtained on graphite substrates, and 1.5% efficient (AM1) devices were made from 50 μm thick Si films on graphite substrates
      Silicon vapor growth is useful in several other types of
structures as well as for thin film applications. In these
structures the substrate is a Si wafer and the Si film is an
epitaxial, single crystal layer. A conventional junction solar
cell can be made by vapor growth of the top region rather than
diffusion; this would eliminate the dead layer region of very
low lifetime and allow the top region to be made thicker for
lower sheet resistance. A second use of Si epitaxial layers
is in the epitaxial BSF device [44,90], where the solar cell
is made on a thin, lightly doped Si layer grown on a heavily
doped Si substrate. This structure magnifies the beneficial
effects of the back surface field on radiation tolerance and
high open circuit voltage, since the active base region is so
thin (∿10 μm). A third use of epitaxial Si is in the novel
vertical multijunction structures described by Smeltzer *et al.*
[134], in which devices similar to Fig. 81b and c were made by
the selective etching of deep slots into <110> oriented Si slice
and eventually epitaxially refilling these grooves with Si to
create the vertical slabs of the multijunction structure. This
is an elegant and relatively reliable method of fabricating thes
multijunction devices, and it may be developed further in the
future, although it is a fairly expensive technique at present.

TABLE 18
Tetrahedral Radii of Impurities in Si[a]

Atom	Radii (Å)	Atom	Radii (Å)
(Si)	(1.176)	(Si)	(1.176)
P	1.07	B	0.91
As	1.18	Al	1.25
Sb	1.35	Ga	1.25

[a]After Rhodes [198] and Celotti *et al.* [199].

2. DOPANTS

There are several dopants that can be used in Si. However, most Si solar cell technologies are based almost entirely on two in particular: boron and phosphorus. Both have drawbacks for use in solar cells. They have relatively poor atomic matches to the Si host lattice, and can therefore introduce strain and dislocations. Table 18 lists the tetrahedral radii of various impurities in Si. Arsenic is seen to be a better match to the lattice than phosphorus, while Al and Ga are better matches than boron. Several authors have attempted to use Al as the base dopant, and reported that slightly greater diffusion lengths and junction perfection are indeed observed compared to boron-doped devices [149], but there have been problems of uniformity in the Al doping in Czochralski-grown Si [150], perhaps due in some way to the high oxygen content [150]. For the n-type diffused regions, phosphorus has been used almost exclusively because of the well-developed diffusion technology, but As should be a superior dopant because of less strain and dislocation generation. Dislocations and other defects introduced by the phosphorus diffusion are believed to be the cause of the very poor lifetimes associated with the dead layer in conventional diffused junctions [4,24].

For space applications, boron and phosphorus also have the undesirable effect of contributing to the radiation degradation by adding to the formation of recombination centers; the higher the concentration of boron in the base of N/P cells, for example, the less radiation tolerance the device has [146, 149]. It is this effect that results in the widespread use of 10 ohm-cm substrates for space applications even though 1 and 0.1 ohm-cm devices have higher predicted initial efficiencies. Aluminum doping as a substitute for boron results in more radiation tolerant cells [149-150], presumably because Al ions do not add to recombination center formation. This should make

it possible to use 1 or 0.1 ohm-cm Al-doped bases to take
advantage of the higher predicted efficiencies.

It has been reported that phosphorus is even worse than
boron in reducing the radiation tolerance in Si solar cells
[200]. There is not enough information at this time to judge
whether As would be better than phosphorus in space, but As
is not likely to be any worse.

Lithium used as the base dopant in P/N cells has the
property of improving the radiation tolerance of solar cells,
as discussed in Chapter 7. When a vacancy is created in the
lattice by an energetic incident particle, an oxygen ion generally combines with it and a recombination center is formed.
If Li is also present in the material, a Li ion can unite with
the oxygen-vacancy defect. If this newly created Li-O-vacancy
point defect were negatively charged, it would have a high
capture cross section for minority carrier holes in the n-type
material and therefore act as a recombination center. Instead
of this, however, the Li-O-vacancy complex is apparently neutral and has a lower capture cross section for minority carriers than the O-vacancy defect alone [201,202], resulting in
a higher lifetime when the Li is present than when it is absent.
It has also been conjectured [201,202] that a second Li ion
can migrate to the defect complex, combine with it, and produce
a center with an even lower capture cross section. The net
result of this behavior is that a Li-doped cell "recovers" from
the initial radiation degradation at a rate determined by the
Li concentration and diffusion coefficient and on the oxygen
content (the greater the oxygen content, the greater is the
number of O-vacancy defects and the slower is the recovery rate)

3.   DIFFUSION

The diffusion process is probably the single most critical
step in the fabrication of solar cells. The temperature, time
duration, and impurity source determine jointly the surface
concentration, junction depth, sheet resistance, and (indirectly)
the lifetime in the diffused region of the cell. For most
solar cells made in the past, phosphorus is diffused into p-type
wafers to obtain an N/P device. Both P_2O_5 and $POCl_3$ have been
used as diffusion sources [200,203-206]; these are transported
to the Si wafer using a carrier gas such as dry oxygen, resulting in the formation of a dopant-containing glass on the Si
surface. Diffusions are then carried out at 800-1000°C for
periods of a few minutes to several hours, resulting in surface
concentrations of around $3-4\times10^{20}$ cm^{-3}, junction depths of
0.2-1.0 μm, and sheet resistivities of around 40-100 ohms/square
Figure 105 shows the diffusion profiles and junction depths

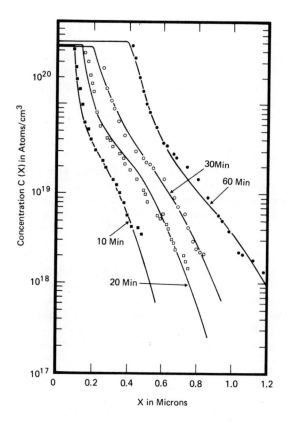

FIG. 105.   Diffusion profiles for 950°C diffusions of phosphorus into Si using a $POCl_3$ source.   Substrate doping $= 2 \times 10^{16}$
atoms/cm^3.   (After Tsai [205]; courtesy of the IEEE.)

for diffusions at 950°C for various periods of time [205].
The electrically active phosphorus concentration remains constant for a distance equal to about a third of the junction
depth, then decays rapidly with distance into the wafer. This
constant concentration region is the so-called dead layer,
characterized by a high density of dislocations [204] and low
lifetime [4,24,35].   Both the stress introduced by the phosphorus ion-Si lattice mismatch and the high concentration of
electrically *inactive* phosphorus atoms [203,204] (as much as
10^{21} cm^{-3}) probably contribute to the poor quality of the Si
in this region.   (With such a high phosphorus concentration,
the material at the surface is not really Si anymore, but a
Si-phosphorus compound or alloy instead.)   Lindmayer and
Allison [4] estimate that the lifetime over at least a portion

of the diffused region could be as low as 100 psec, and measured, average values of lifetime over the entire diffused
region of several nanoseconds have been reported [24]. The
most important step in eliminating the dead layer is to reduce
the surface concentration and junction depth [4], which raises
the lifetime and minimizes both surface recombination and bulk
recombination losses by providing an electric drift field.
(The higher sheet resistances due to the reduced doping and
width of the diffused region must be compensated for by more
contact "fingers.")

Kamins [207] has described a method for growing doped
oxides on Si to use as diffusion sources. Silane mixed with
phosphine or diborane can be passed over Si wafers in an oxi-
dizing atmosphere, resulting in SiO_2 with a concentration of
boron or phosphorus determined by the gas mixtures. Surface
concentrations in the $5\text{-}8\times10^{19}$ cm^{-3} range were easily obtained.

Boron diffusions have been receiving increased attention
lately because of the success of the Li-doped P/N solar cells,
where Li forms the n-type dopant for the base and boron is
used for the diffused top region. The mismatch between boron
ions and the Si lattice is capable of introducing stress and
dislocations [82] in an analogous manner to the phosphorus
diffusions, and an equivalent dead layer adjacent to the surface is probably obtained in many cases. The most common
boron diffusion technique for solar cells is to use BCl_3 or
BBr_3 in an oxygen atmosphere to deposit a boron compound on
the Si surface, followed by diffusing at temperatures of around
900-950°C for 5-15 min. Either the use of water vapor in place
of oxygen [208] or the silane-diborane-oxygen method of Kamins
[207] can be used to lower the boron surface concentration
(normally 2×10^{20} cm^{-3}) to $5\text{-}8\times10^{19}$ cm^{-3} and minimize the dead
layer.

Lithium incorporation in the base of P/N cells to obtain
high radiation tolerance is performed after the boron diffusion
has been completed. The Li is applied [82] to the back of the
substrate either by painting on an oil solution of Li powder
or by direct Li evaporation, and the Li is diffused into the
Si at 340-400°C for 2-8 hr, with the most reliable results obtained when the resulting Li concentration is in the range of
$5\times10^{14}\text{-}1\times10^{16}$ cm^{-3} averaged over the base, with a concentration
gradient (measured by capacitance-voltage techniques) at the
junction edge of $5\times10^{18}\text{-}4\times10^{19}$ cm^{-4}.

It may be possible, especially for terrestrial applications,
to use "paint-on" diffusion sources such as B- and P-doped SiO_2
solutions which are spun-on similarly to photoresist. The
reliability of these sources might be less than conventional
techniques, but the cost of the diffusion step might be significantly lowered in this way.

B. GaAs Solar Cells

1. CRYSTAL GROWTH

Techniques for producing GaAs wafers for making solar cells are very similar to those for producing Si crystals, except that care must be taken to ensure that stoichiometry (equality in the number of Ga and As ions) is maintained. GaAs ingots have been grown by Czochralski [209], Bridgman [210], and ribbon techniques [211,212], but the method most often used is the horizontal Bridgman technique [210]. In this method, a Ga-containing boat is placed at one end of a sealed quartz tube and a source of As is positioned at the other end. The Ga is held at a temperature just above the GaAs melting point (1240°C), and the As is held at around 615°C where the As vapor pressure is one atmosphere. Arsenic is transported by diffusion from the source region to the Ga melt, which eventually becomes converted into molten GaAs, and since the As vapor pressure over molten GaAs at 1240°C is also 1 atm, nearly perfect stoichiometry is achieved. The boat of liquid GaAs, with or without a crystal seed at one end, is then drawn slowly through the temperature gradient between the two temperature zones at a rate of around 1 cm/hr, and solidification of the single crystal ingot takes place. Growth by the Czochralski method is almost identical, except that a crystal seed is dipped into the molten GaAs and pulled slowly out of the melt in the vertical direction while rotating to enhance uniformity.

The growth of GaAs ribbon is still mostly in the planning stage. The method proposed [212] involves encapsulation of the GaAs during the growth process to prevent volatilization at the needed high temperatures. It is possible that the EFG technique could be used, but it would have to be in a closed system to maintain the As pressure at 1 atm to ensure stoichiometry. The cost advantage obtained with ribbon Si would not be obtained with ribbon GaAs of comparable thickness, however, since the cost of the starting material in GaAs devices is a larger part of the cost of the finished device than in Si.

GaAs vapor growth has been a well-developed technology for a number of years and should be directly applicable for the growth of thin polycrystalline or single crystal films for terrestrial applications. The most common method involves the transport of Ga using HCl or $AsCl_3$ diluted with hydrogen and the mixture of the gallium chloride with arsine or another arsenic-containing species in the vicinity of a suitable substrate at temperatures sufficient for the chemical reactions resulting in the deposition of GaAs to take place [213,214].

Layers can be grown at temperatures of 600-800°C at rates of
up to 10 µm/hr, and since only very thin layers are needed to
produce highly efficient devices (Chapter 5), the time and
energy consumed in the process can be relatively small.
    Organometallic compounds (e.g., gallium trimethyl) can
be used in the growth of both GaAs [215,216] and $Ga_{1-x}Al_xAs$
[217] layers, but these compounds are relatively expensive
at the present time.
    The most successful method for producing GaAs and
$Ga_{1-x}Al_xAs$ for high efficiency devices is the LPE technique
[30,32,218,219]. The high quality (in terms of lifetime,
mobility, and freedom from defects) of the LPE materials derives from growth under near equilibrium conditions and the
cleansing action of the Ga by which impurities are retained
in the liquid rather than being incorporated in the growing
crystal. The technique involves saturating a melt of Ga with
GaAs at temperatures of around 900°C, bringing this saturated
solution in contact with a GaAs (or other) substrate, cooling
over a specified temperature range at 0.1-0.5°C/min to obtain
the epitaxial layer, and decoupling the melt and substrate to
prevent further growth when the desired thickness has been
obtained. The dopant can be added directly to the melt and
will be incorporated in the growth in accordance with the
segregation coefficient. For p-type GaAs, Ge is the dopant
most commonly used; for n-type, Sn or Te are used.

2. DOPANTS

    The dopants for GaAs fall into several categories, vola-
tile and nonvolatile, amphoteric and nonamphoteric. For
volatile impurities such as S, Se, Te, Cd, and Zn, significant
amounts of dopant may be lost during the crystal growth due
to the high vapor pressure of the dopant at the growth temperature, and the amount and uniformity of the dopant incorporated
into the crystal depends not only on the distribution coefficient but on the vapor pressure of the impurity as well (the
same is true for phosphorus doping of Si ingots, but not for
boron doping, since boron has a negligible vapor pressure).
Less volatile dopants such as Ge, Si, Mg, and Sn can be added
directly to the melt in the desired amount without much concern
for the vapor pressure. The distribution coefficients for
these impurities in GaAs, as given by Willardson and Allred
[220], are shown in Table 19, and the tetrahedral radii in
Table 20. Since the radii of As and Ga are 1.18 and 1.25 Å,
respectively, the least amount of strain and defect generation
might be expected for impurities within this range, and there

TABLE 19
Distribution Coefficients of Impurities in GaAs[a]

Impurity	Coefficient	Impurity	Coefficient
S	0.3	Zn	0.4
Se	0.3	Si	0.14
Sn	0.08	Mg	0.1
Te	0.059	Ge	0.01

[a]After Willardson and Allred [220].

does seem to be some correlation of radii with material properties; Ge as a p-type dopant, for example, which matches the
GaAs lattice quite well, appears to yield consistently higher
lifetimes and minority carrier diffusion lengths than other
p-type dopants.

Si, Ge, and Sn are amphoteric dopants; they behave as
donors if they occupy the Ga site and acceptors if they occupy
the As site. For stoichiometric crystals, the growth temperature and impurity concentration largely determine which site
is occupied. At the GaAs melting point, Si substitutes for
Ga and produces n-type material, while Ge has an equal tendency
for both sites and produces compensated crystals. Below about
900°C (as used in LPE growth), both Si and Ge occupy the As
site and behave as acceptors. Tin apparently produces n-type
material under all conditions.

3. DIFFUSION

Most GaAs solar cells have been made by diffusion, and
since the diffusion coefficient of Zn in GaAs is considerably
greater than that of any other shallow dopant, nearly all the
diffused GaAs cells have been of the P/N variety made by Zn
diffusion into n-type substrates [6,38,221,222], although Cd
diffusion was also attempted in the early days [222]. Casey
[223] has summarized the theory and experimental results for
the diffusion of Zn and other dopants into GaAs and other
III-V compounds.

Since GaAs solar cells are strongly dominated by surface
recombination and low lifetimes in the diffused region, it is
important to ensure that the junction depth is small (<0.5 μm)
while at the same time the surface concentration is high to
prevent significant sheet and contact resistances. The high
diffusion coefficient of Zn and the temperature dependence of

TABLE 20
Tetrahedral Radii of Impurities in GaAs[a]

Atom	Radii (Å)	Atom	Radii (Å)
S	1.02	Zn	1.30
Se	1.16	Si	1.17
Sn	1.40	Mg	1.36
Te	1.45	Ge	1.22

[a]After Willardson and Allred [220].

the surface concentration make it difficult to satisfy these
two criteria simultaneously. In the most common diffusion
method, an n-type GaAs wafer is placed in an evacuated quartz
ampoule along with a source of Zn, which can be elemental Zn,
a Zn-Ga mixture, a Zn-As mixture, or a Zn-Ga-As combination.
The sealed ampoule is then placed in a furnace at 600-800°C
for a few minutes to an hour, depending on the source and on
the depth desired. Elemental Zn and Zn-Ga mixtures tend to
produce considerably deeper junctions than the other two sources,
while the Zn-Ga-As combination yields shallower junctions with
higher surface concentrations and improved reliability. For
temperatures above 700°C, it is necessary to provide an As over-
pressure in the ampoule (e.g., by the use of Zn-As combinations)
to prevent any dissociation of the GaAs.

The surface concentrations as a function of temperature
for diffusions using an elemental Zn source are shown in Table
21, along with the Zn diffusion coefficient at that temperature
and dopant concentration. For these conditions, junction
depths of about a micron are obtained in 10 min at 700°C and
in less than a minute at 800°C, illustrating the difficulty
in obtaining small x_j's with high C_0's. Tsaur *et al.* [38]
have proposed diffusing at 600-650°C to obtain shallow junctions
at reasonable times, even though the surface concentrations
would be lower than for diffusions at higher temperatures.
Casey and Panish [223] have shown that a mixed source of 5/50/45
at.% of Ga/As/Zn yields higher Zn surface concentrations and
smaller junction depths than an elemental Zn source for diffusions at 650-700°C; concentrations of 2×10^{20} cm^{-3} were obtained
at 700°C using this mixture. Marinace [224] has developed a
diffusion technique using a Zn-saturated InAs source together
with a small amount of Cd_3As_2 in the ampoule; the InAs-Zn
source yields low Zn concentrations and shallow junctions and
the Cd yields a high acceptor concentration (1×10^{20} cm^{-3}) at
the GaAs surface.

TABLE 21
Surface Concentrations of Zn Using
an Elemental Zn Source[a]

T(°C)	$N_{Zn}(cm^{-3})$	$D(cm^2/sec)$
700	8.5×10^{19}	2×10^{-10}
800	1.6×10^{20}	3×10^{-9}
900	2.8×10^{20}	3×10^{-8}
1000	4×10^{20}	7×10^{-8}

[a]After Casey [223].

4. $Ga_{1-x}Al_xAs$-GaAs DEVICES

The difficulties of overcoming surface recombination and
low lifetime in the diffused region are largely overcome by
the addition of a $Ga_{1-x}Al_xAs$ window. The $pGa_{1-x}Al_xAs$-pGaAs-
nGaAs solar cells [8,9,36,45,84,225] are fabricated by LPE.
A melt consisting of Ga, Al, Zn, and GaAs is brought into contact with an n-type GaAs substrate and a layer of Zn-doped
$Ga_{1-x}Al_xAs$ is grown by cooling for a few degrees at 0.1-0.5°C/
min. During this process, which takes place around 900°C, Zn
also diffuses into the GaAs substrate and forms a p-n junction
in it, but the junction is relatively shallow (0.5-3.0 μm)
compared to ampoule diffusions at this temperature due to a
much lower Zn-surface concentration ($\sim 10^{18}$ cm^{-3}). The $Ga_{1-x}Al_xAs$
layers are 1-10 μm in thickness, with Al compositions of 70-
90%; these conditions lead to high predicted [45] and measured
[8-9] efficiencies for both outer space and terrestrial sunlight.
    Other dopants, such as Mg and Ge, can be used for the
pGaAs and $pGa_{1-x}Al_xAs$ regions. These dopants diffuse more
slowly than Zn, however, so that diffusion of the pGaAs region
may be more difficult, particularly for Ge. In this case an
epitaxial pGaAs layer can be grown prior to growing the epitaxial $Ga_{1-x}Al_xAs$ layer.

C. Cadmium Sulfide Solar Cells

The very large interest in CdS solar cells arises from
their potentially high efficiency per unit cost. Their efficiency at AM1 is at best around 10% (although in theory as much
as 15%), but their cost could potentially be less than the
ultimate cost of Si cells for terrestrial applications, leading
to visions of both small, rooftop power systems and large,
kilomegawatt systems.

The low cost of CdS cells arises from the use of evaporated thin films of the material, rather than single crystals, and from the simple processing steps that allow them to be fabricated in large areas on a near production line basis. A sheet of Cu or Mo 1-2 mil thick or a sheet of metallized plastic such as aluminized Kapton is used as the substrate. A CdS polycrystalline film up to about 1 mil thick is evaporated onto the substrate sheet, and the surface is lightly etched to obtain as clean a condition as possible. The unit is then dipped into a hot (80-100°C) cuprous chloride solution for 10-30 sec to obtain a copper sulfide (Cu_xS) layer between 1000 and 5000 Å thick on the CdS surface, with the Cu_xS thickness depending on both the solution concentration and the dipping time [47]. Contact to the Cu_xS is made with a Cu or Au grid either laid or plated onto the surface, and the surface is laminated with a 1 mil thick Kapton, Aclar, or Mylar sheet held on with transparent adhesive. The units are usually given a 2-5 min heat treatment at 250°C to improve the rectifying properties and the photosensitivity of the Cu_xS-CdS heterojunction. The resulting solar cells are usually 50-60 cm^2 in area and have an operating voltage and current of around 0.33 V and 0.8-0.9 A, respectively.

Historically, CdS cells have been plagued with a number of degradation problems. These cells have been known to degrade under a variety of environmental conditions: (1) under high humidity; (2) at high temperatures (>60°C) in air; (3) at high temperatures when illuminated; (4) when the load voltage exceeds 0.33 V; or (5) after temperature cycling (-150°C to +60°C) for a number of times.

Water vapor causes a decrease in I_{sc} to occur, but leaves V_{oc} and FF largely unchanged [226]. Moisture is capable of penetrating the plastic lamination and becoming absorbed into the underlying Cu_xS-CdS structure, resulting in electronic traps which lower the collection efficiency. The initial I_{sc} can be recovered by heating the cell at 180°C in vacuum for several hours. A second, irreversible degradation can occur if the plastic or adhesive absorbs the moisture; Mylar sheets with epoxy adhesive are superior to Kapton with Capran adhesive in this regard [226]. Careful, thorough encapsulation with nonporous materials should largely eliminate the humidity problem.

Heating a cell in *vacuum* in the dark to temperatures as high as 200°C has no appreciable effect on it, but if a device is taken much above 60°C *in air,* irreversible decreases in I_{sc} can occur [227,228], which are attributed to oxygen and moisture attack of the Cu_xS and its conversion to mixtures of CuO and Cu_2O As with the humidity problem, more thorough encapsulation should be capable of minimizing the amount of air reaching the Cu_xS layer. A second degradation of I_{SC} can take place at temperatures above 60°C when the device is *illuminated*, even if no air is present. This degradation is apparently caused by a light-activated phase change in the Cu_xS, i.e., Cu_2S goes to lower forms of Cu_xS [54,228,229]. Experimentally, it appears that even slight deviations of the Cu_xS from perfect stoichiometry (x = 2) lowers the efficiency considerably. Bogus and Mattes [54] have greatly improved the high temperature stability of CdS cells by evaporating a 100 Å layer of Cu onto the Cu_xS surface and driving it in by subsequent heat treatment, ensuring that the Cu_xS remains nearly stoichiometric Cu_2S. This Cu evaporation step also helps to mask against the oxygen attack mentioned above, retarding the conversion of Cu_xS to Cu_xO.

If a CdS cell with light incident is operated at a load voltage greater than 0.33-0.35 V, degradation in both the V_{OC} and FF can take place, while I_{SC} remains largely unchanged [54,228-230]. This behavior is accompanied by the appearance of metallic copper within the Cu_xS layer and near the interface. Bernatowicz and Brandhorst [229], Mathieu *et al.* [231], and others have shown that a light-activated electrochemical reaction takes place at a threshold of 0.35 V, at which Cu_2S converts into CuS and Cu. The released Cu ions form fine filaments that act as shunt paths across the junction, degrading the electrical performance. This voltage-induced degradation can be minimized by ensuring that the load voltage never exceeds 0.33 V, but more importantly, it appears that it can be eliminated altogether by doping the CdS to 5 ohm-cm or below [232], which seems to inhibit the movement and precipitation of Cu across the Cu_xS-CdS interface. The transport of Cu across this interface appears to be the single most damaging cause of degradation in these devices [10], resulting in both a decrease in the short circuit current under some conditions [55,78] and a reduction in voltage output and FF when filaments are formed. The prevention of these problems by doping the CdS is a major breakthrough in solving the instability problems of CdS solar cells.

Finally, thermal cycling and the resulting stresses within the cell due to the difference in expansion coefficients between the plastic and the two semiconductors can cause delamination of the plastic encapsulation and a lifting of the stripe contacts to the Cu_xS, with consequent loss of I_{SC} and increased series resistance [226,229]. This problem is not severe for devices operated at the earth's surface, since the temperature range on earth is not very large, but devices in space, where temperatures can rise to 80°C or higher in sunlight and drop

to -150°C or lower in the dark, are likely to fail after a
long enough period of time. Electroplated Au contact stripes
in place of the pressure-applied Cu grids can reduce the problem considerably, and it would seem that other types of plas-
tics and adhesives could be found that would be better from
the thermal expansion point of view.

Overall, then, it would seem that considerable progress
has been made in understanding and overcoming the instability
problems of Cu_xS-CdS solar cells [232a], and if they can be
made reliable enough in a practical structure, they could have
a major impact on terrestrial energy needs. Experimental CdS
solar cell panels have already been placed on houses and have
performed well for several years [11], using a dry N_2 atmosphere to prevent water vapor and O_2 from reaching the cells,
and cooling the back of the devices with an air flow to keep
the temperature below 60°C [10,11] (the devices available at
the beginning of this experiment did not have all the instability
correcting features). Using accelerated life-test studies,
estimates of 20 yr or more for the useful life of carefully
protected CdS solar cells have been made [10,11], and estimates
of the cost of producing large amounts of electrical power using
these cells have ranged from less than $100/av kW [233,234]
to about $500/av kW [11], about the same as the present cost
of a conventional fossil fuel power plant. (Estimates of the
cost of generating electricity using ribbon or thin film Si or
thin film GaAs fall in the same ball park.)

Jordan [234] has described an intriguing method of mass-
producing CdS solar cells on a large-scale, continuous basis.
The method makes use of the existing glass sheet technology.
Large glass sheets are produced in continuous fashion using
the float-glass process [234], where raw materials are fed in
one end of a furnace and molten glass is poured out the other
end onto a bed of molten tin. Thin films of the transparent
conductor SnO_x are sprayed onto the glass surface, followed
by deposition of 2-5 μm thick CdS films, also by spraying.
Other chemicals are sprayed onto the CdS surface to produce
the Cu_2S layer after the glass-SnO_x-CdS sheets have cooled
below 150°C. The device is finished by cutting the sheets to
the desired size and evaporating the top electrodes. These
cells are illuminated through the back, and the transparent
SnO_x acts as the back electrode. The whole process can easily
be automated, and cost projections are very low ($52/peak kW,
$250/av kW for the entire power generating plant). Prelimi-
nary devices have already been made by a prototype process
[234], but little data on the process or the resulting devices
are available as yet.

D. Heterojunctions and Schottky Barriers

The most promising heterojunction pairs from the lattice match, electron affinity, and expected efficiency points of view are $Ga_{1-x}Al_xAs$-GaAs, ZnSe-GaAs, GaP-Si, and ZnS-Si.

The $pGa_{1-x}Al_xAs$-nGaAs device is similar to the three-layer structure already mentioned (Section B4) except for the absence of a pGaAs region. Solar cells of this type have been made by LPE [7], but vapor growth should also be readily adaptable [217]. If a true heterojunction is desired, care must be taken to prevent the formation of a pGaAs region while the $Ga_{1-x}Al_xAs$ layer is being grown. If Zn is used as the acceptor species, the nGaAs substrate must be doped above 10^{18} cm^{-3} to inhibit the pGaAs layer from forming. The use of Ge to dope the $Ga_{1-x}Al_xAs$ should obviate the problem because of the low diffusion coefficient of Ge in GaAs at 900°C and below.

nZnSe-pGaAs heterojunctions have been grown by both LPE [235] of GaAs on ZnSe from Sn solution at 520-560°C and by vapor growth [236,237] of ZnSe on GaAs substrates at 490-610°C. The conversion efficiency of the LPE device was low (1%) [235] because of the very high doping level in the GaAs ($>10^{19}$ cm^{-3}) and because of high series resistance due to the high resistivity of the ZnSe (>1 ohm-cm), but the good spectral response at high photon energies verified the theory that the interface recombination velocity should be low in this heterojunction. No electrical-optical measurements were reported on the vapor grown devices, but if the difficulties of obtaining low resistivity ZnSe can be overcome, it should be possible to make highly efficient solar cells between ZnSe and GaAs.

Epitaxial ZnS-Si and GaP-Si heterojunctions have been grown by a number of methods. Single-crystal ZnS has been grown on Si [238] by H_2 transport at temperatures of 450-600°C, with growth rates of 1300 Å/hr or less. The thin oxide which is invariably present on Si surfaces was removed just before growth by a vapor etch in H_2 at 1250°C followed by etching in HCl at the same temperature. GaP-Si devices have been made by evaporation [239], vapor transport with HCl [240], vapor synthesis using organometallic compounds [241], and LPE [242]. In all cases, steps such as those mentioned were taken to eliminate the unwanted oxide from the Si surface just before growth. Temperatures of 750-1150°C were used, but single crystal layers were obtained only above 900°C. At temperatures above 1000°C, very high growth rates could be achieved, as much as a micron per minute. The major problem encountered was cracking of the GaP layers when the devices were cooled from the growth temperature; the cracking is due to the stress caused by the thermal expansion difference between GaP and Si

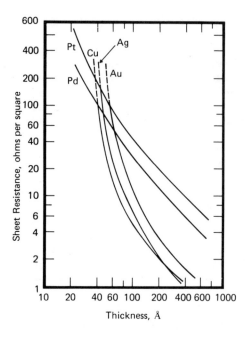

FIG. 106. Sheet resistances
of thin metal films. (After
Stirn and Yeh [107]; courtesy
of the IEEE.)

(the lattice match of this pair is good). Slower cooling rates
and thinner GaP epitaxial layers were helpful in reducing the
cracking problem. No solar cells have been reported using
GaP-Si heterojunctions as yet.

Schottky barrier solar cells are probably the simplest
of all types to fabricate, requiring only an Ohmic contact at
the back and a semitransparent metal at the front, along with
the usual contact grid pattern to lower the series resistance.
The transparent metal film is normally evaporated onto the
carefully prepared semiconductor surface [103,107], and films
of about 100 Å thickness yield transmissions of around 60%
with sheet resistivities of 5-50 ohms/square. Figures 70 and
71 showed the transmission through thin gold films as a function of thickness and wavelength, respectively, and showed
that the addition of a proper antireflection coating can reduce
the optical loss due to the metal film down to a few percent
[106]. Figure 106 shows the sheet resistances of various metal
films as a function of thickness, as presented by Stirn and
Yeh [107]. Gold and silver appear to be good prospects for
Schottky barrier solar cells in terms of low sheet resistance,
good barrier heights (Table 8), and high transparency. Plati-
num is probably even better, but in the past has been more
costly and more difficult to work with.

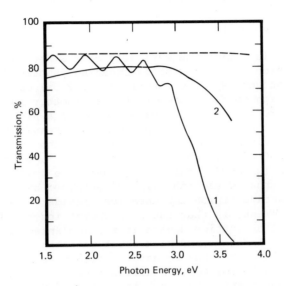

FIG. 107. Transmission of light through transparent conducting layers on sapphire substrates: (1) In₂O₃/SnO₂; (2) SnO₂/Al₂O₃ wafer. (Reflection accounts for most of the loss at low energies.)

One of the potentially most valuable types of heterojunction or Schottky barrier solar cells has received very little attention in the past. There is a class of materials known as transparent conducting glasses, consisting of In_2O_3, SnO_x, ZnO, and the like, which have bandgaps of 3 eV or above and resistivities of 0.0005 ohm-cm or less under some conditions. Such materials could possibly be used to make heterojunction devices with p-type semiconductors or Schottky barriers with n-type semiconductors. The transparency of these conductors is high (Fig. 107), usually above 90% for layers several thousand angstroms to a micron in thickness, and their sheet resistivities can be as low as several ohms per square [243]. They can be evaporated [244], sprayed [234,245], or sputtered [243, 246] onto a suitable substrate, and are potentially economical in both cost and energy consumption. Several solar cell devices of both the Schottky barrier type [247] and heterojunction type [248] have been made using SnO₂ on Si and GaAs substrates, but the efficiencies were low (∼1%), probably due to the poor quality of the SnO₂ films or to detrimental energy barriers at the interface. Efficiencies in excess of 10% should be obtainable with very low resistivity transparent coatings on Si and GaAs.

E.   Ion Implantation

Ion implantation theoretically offers several advantages
over diffusion as a means of fabricating junctions. One advantage is that very shallow junctions (<0.2 µm) can be made even
with very high surface concentrations; another is that the
implanted profile can be adjusted to yield a high electric
drift field throughout the top region. (Presumably, diffusion
should also yield such a drift field, but the concentration
dependence of the diffusion coefficient of many impurities in
Si and GaAs [35,249] tends to eliminate the drift field near
the surface that would otherwise be obtained by the diffusion.)
The disadvantage of ion implantation, beside the capital cost
of the equipment, is that the high energy ion species creates
a great deal of lattice damage in the implanted region, resulting in a low lifetime and high sheet resistance. High temperature annealing must then be used to eliminate the damage and
restore the material to crystalline form.

Ion implantation has been investigated for both Si [143,
250] and GaAs [251-252] solar cells. The p-type Si substrates
were implanted with phosphorus to a depth of about 1 µm, with
subsequent annealing at 500-650°C [250]. Very high short circuit currents ($\geq$40 mA/cm^2) were obtained for AM0 conditions,
but the open circuit voltages were low, less than 0.5 V even
after annealing. The highest efficiency was around 8%. The
n-type GaAs substrates were implanted with Zn [251] or Be [251,
252]. Beryllium was preferred over Zn as an implanting species
because the lower mass of the Be ion allowed implantation at
lower accelerating voltages with consequent reduced lattice
damage [251]. Beryllium also resulted in lower sheet resistivities compared to Zn for the same dose. Annealing of the Be-
implanted GaAs samples was carried out at 500-800°C. Annealing
at 600°C was sufficient to obtain utilization factors (# ions
electrically active/# ions implanted) of 80% [251]. The FF
and open circuit voltage improved continually as the annealing
temperature was raised [252], but the maximum open circuit
voltages obtained were about 0.8 V after a 700-800°C anneal
[252], considerably less than the open circuit voltages usually
obtained from devices made by diffusion. The short circuit
current of the Be-implanted cells was also much lower than
expected.

The low open circuit voltages of both Si and GaAs implanted
devices compared to diffused or epitaxial structures at the
same doping levels is reminiscent of the low output voltages
obtained after low energy proton irradiation, suggesting that
not all the damage near the junction has been removed under
the annealing techniques used so far. Unless the low output

TABLE 22

Refractive Index of Si, GaAs, 300°K[a]

λ (μm)	n_{Si}	n_{GaAs}
1.1	3.5	3.46
1.0	3.5	3.5
0.90	3.6	3.6
0.80	3.65	3.62
0.70	3.75	3.65
0.60	3.9	3.85
0.50	4.25	4.4
0.45	4.75	4.8
0.40	6.0	4.15

[a]After Kirk-Othmer [253] and Willardson-Beer [254].

voltages can be improved, they pose a serious restriction to the use of ion implantation as a technique of fabricating solar cells.

F. Antireflective Coatings

The antireflective (AR) coating is one of the most important parts of a solar cell design. Materials such as Si and GaAs have high indices of refraction (Table 22). For Si, the loss of incident light amounts to 34% at long wavelengths (1.1 μm) and rises to 54% at short wavelengths (0.4 μm) [255, 256]. A proper single layer AR coating can reduce the reflection to 10% averaged over this wavelength range ($\sim$8% with the addition of a coverslip), and a double layer coating can reduce it to around 3% on the average.

The total reflection of incident light of wavelength λ from the surface of a material covered by a single nonabsorbing coating of thickness d is given by [256,257]

$$R = (r_1^2 + r_2^2 + 2r_1r_2 \cos 2\theta)/(1 + r_1^2r_2^2 + 2r_1r_2 \cos 2\theta) \qquad (112)$$

where r_1 and r_2 are the individual reflectances

$$r_1 = (n_0 - n_1)/(n_0 + n_1), \qquad r_2 = (n_1 - n_2)/(n_1 + n_2), \qquad (113)$$

and θ is the phase thickness of the optical coating

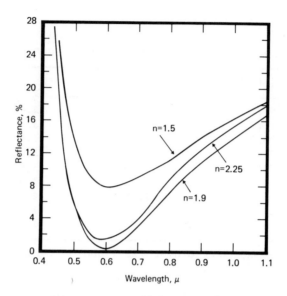

FIG. 108.   Calculated reflectances of single layer antireflective coatings on Si, for three values of refractive index.

$$\theta = 2\pi n_1 d_1 / \lambda, \tag{114}$$

and n_1 is the refractive index of the coating, d_1 is its thickness, and n_2 is the refractive index of the underlying semiconductor. The reflectivity has a minimum at a quarter wavelength, where $n_1 d_1 = \lambda_0/4$, and for odd multiples of $\lambda_0/4$. This minimum is given by

$$R_{min} = \left[(n_1^2 - n_0 n_2) / (n_1^2 + n_0 n_2) \right]^2 \qquad [\lambda = \lambda_0] \tag{115}$$

and is equal to zero if $n_1 = n_0 n_2$. Since n_0 is equal to 1 for an air medium, the refractive index of the film should equal the square root of the index of the semiconductor for zero reflectance. The reflection is higher for wavelengths either higher or lower than this quarter wavelength value due to the cosine function and due to variations in the refractive indices with wavelength. Figure 108 shows the calculated reflection from a Si surface for AR layers with indices of 1.5, 1.9, and 2.25, representing SiO_2, SiO or Al_2O_3, and TiO_2 or Ta_2O_5, respectively. The AR layer having an index closest to $\sqrt{n_{Si}}$ results in the lowest reflectance, but higher index coatings result in only slightly larger reflectances. Low index single layer coatings on the other hand result in considerably higher reflectances, and should be avoided whenever possible.

The material most often used for AR coatings in the past has been SiO. This material has an index of 1.8-1.9 and results in a very low reflectance minimum (<1%), but it does have some absorption loss in the visible region. CeO_2 has been suggested [258] as a substitute for SiO, since it has less absorption, but the processing of CeO_2 is more difficult than SiO [258], making it less desirable from a fabrication point of view, aside from the fact that it also has a somewhat higher index. Al_2O_3, with an index of 1.86, should be a good substitute for SiO, as should Si_3N_4.

Lower average reflectivity can be obtained [256,259,260] using two antireflection coatings instead of one, where the first layer, next to the semiconductor, has an index of 2.2-2.6 and the top layer has an index of 1.3-1.6. The double layer system is a better "impedance match" between the high index of the semiconductor and the low index of air. The reflection of light from the surface of the overall device is given by [256]

$$R = \frac{\left(\begin{array}{c} r_1^2 + r_2^2 + r_3^2 + r_1^2 r_2^2 r_3^2 + 2r_1 r_2(1+r_3^2)\cos2\theta_1 + 2r_2 r_3(1+r_1^2)\cos2\theta_2 \\ + 2r_1 r_3 \cos2(\theta_1+\theta_2) + 2r_1 r_2^2 r_3 \cos2(\theta_1-\theta_2) \end{array}\right)}{\left(\begin{array}{c} 1 + r_1^2 r_2^2 + r_1^2 r_3^2 + r_2^2 r_3^2 + 2r_1 r_2(1+r_3^2)\cos2\theta_1 + 2r_2 r_3(1+r_1^2)\cos2\theta_2 \\ + 2r_1 r_3 \cos2(\theta_1+\theta_2) + 2r_1 r_2^2 r_3 \cos2(\theta_1-\theta_2) \end{array}\right)} \quad (116)$$

where r_1, r_2 were given by (113) and

$$r_3 = (n_2-n_3)/(n_2+n_3) \tag{117}$$

where n_3 is now the index of the semiconductor and θ_1, θ_2 are the phase thicknesses of the two coatings

$$\theta_1 = 2\pi n_1 d_1/\lambda, \qquad \theta_2 = 2\pi n_2 d_2/\lambda. \tag{118}$$

The reflectance has either a minimum or a local maximum for quarter wavelength optical coatings ($n_1 d_1 = n_2 d_2 = \lambda_0/4$). This reflectance is given by [256]

$$R = [(n_1^2 n_3 - n_2^2 n_0)/(n_1^2 n_3 + n_2^2 n_0)]^2 \qquad [\lambda = \lambda_0], \tag{119}$$

which approaches zero if the condition $n_2^2/n_1^2 = n_3/n_0$ is fulfilled, and approaches a local maximum with zero reflectance on either side if the condition $n_1 n_2 = n_0 n_3$ is fulfilled. In either case, the average reflectance is lower over a broader wavelength range than for a single layer coating. Figure 109

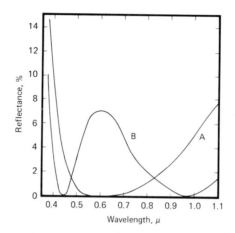

FIG. 109.   *Reflectance of double layer coatings on high index substrates (n = 3.45) such as Si.   A: n_1 = 1.38; n_2 = 2.56. B: n_1 = 1.56; n_2 = 2.21.   (After Musset and Thelen [260].)*

shows the reflectance as a function of wavelength for two types of double layer coatings on Si, as calculated by Musset and Thelen [260].   In contrast to the cosine shape of the single layer coatings (Fig. 108), double layer coatings have either a "U" or a "W" shape with fairly low reflectance over a wide spectral range.   A system consisting of about 600 Å of TiO_2 (n = 2.3) and 1050-1100 Å of SiO_2 or MgF_2 (n $\approx$ 1.4), for example, can reduce the reflection to around 3% averaged over the entire Si response range.

A cover glass used to minimize radiation degradation of solar cells also helps to lower the amount of light lost by reflection, since the cover glass and its adhesive, with their fairly low indices of refraction (1.3-1.5), can act as part of the antireflection coating system [169,180,258,259]. TiO_2 layers are theoretically better than SiO after the cover glass is added, since TiO_2 has both a higher index and less absorption than SiO [4] throughout the visible region.   Both of these materials, however, begin to absorb light strongly for wavelengths shorter than 4000 Å.   For the violet cell, with its high spectral response at short wavelengths, it was necessary to develop a coating material with greater transparency at short wavelengths.   Tantalum oxide [4,261,262] seems to fill this need very well, having a bandgap of around 4.2 eV and a refractive index of 2.20-2.26, and the very high AMO efficiencies of the violet cell [4] have been obtained with Ta_2O_5 coatings and quartz cover glasses.

The transparency and reflectivity of materials such as SiO, TiO_2, and Ta_2O_5 can be strongly affected by the conditions under which the films are obtained. SiO and TiO_2 layers are usually applied by evaporation, and the properties of these films are influenced by the evaporation rate and substrate temperature [263] and by the oxygen pressure in the evaporation chamber (TiO_2 must be evaporated in the presence of excess oxygen). As a general rule, low deposition rates (1-5 Å/sec) and moderately high substrate temperatures (100-250°C) result in highly transparent films with low indices, while higher rates and lower temperatures increase both the absorption in the films and their indices, probably due to a higher concentration of defects.

Antireflective layers can be obtained by evaporation, sputtering, anodization, chemical vapor deposition, and even spinning (similarly to photoresist). Revesz [262] has pointed out that methods such as anodization and chemical deposition (vapor growth) will generally lead to higher quality AR films because they result in well-defined noncrystalline layers with short range order. Noncrystallinity is important to prevent scattering at grain boundaries which decreases the transparency. The short range order is valuable in reducing absorption that occurs in the absence of this order [262]. The Ta_2O_5 used by Comsat Laboratories as the AR coating for the violet cell is produced by chemical vapor growth (method undisclosed), which results in films with the best optical properties. By way of contrast, sputtered Ta_2O_5 layers were reported to show considerable absorption in the visible region [262].

As an alternative to the application of AR coatings, the reflection of light from a semiconductor surface can be minimized by preparing the surface in the form of myriads of pyramids or whiskers with dimensions and spacings of the order of microns. Light incident on this serrated surface can undergo several reflections, increasing the probability of its absorption. The light enters the material at an oblique angle, and absorption of the longer wavelength light therefore occurs closer to the junction instead of deep in the base where the carriers have a greater chance of being lost. Comsat Corporation has announced a 15% efficient Si solar cell made in this manner [264].

G. Ohmic Contacts

The Ohmic contacts to the top and bottom of a solar cell are another important part of the overall device, since any significant contact resistance adds to the series resistance

and reduces the power output.   In addition to low contact
resistance, other desirable features for solar cell contacts
include high bond strength under mechanical stress and the
ability to withstand temperature cycling [52,180].   The upper
contact grid must also be designed properly to allow in the
maximum amount of light while keeping the series resistance
at acceptably low values.

Considerable work has been expended on contact grid design.
The grid found on most commercial cells today consists of a
single bar running the length of the cell with six fingers
coming off at right angles (Fig. 1).   This design is marginally
adequate for conventional N/P or P/N Si devices, resulting in
0.2-0.25 ohm series resistance for a 4 cm^2 cell and a 5-8%
loss of power output under AM0 conditions.   Increasing the
number of contact grid lines lowers R_s considerably [180],
which is essential for operating devices at more than 1 solar
intensity.   The violet cell with its reduced junction depth
and lower doping level in the diffused region uses a contact
structure consisting of 60 lines instead of the usual 6, and
series resistances of 0.05 ohm for a 4 cm^2 device are obtained
[4] even though the sheet resistance of the diffused layer is
500 ohm/square or higher.   Figure 110a shows a thin film of
material of thickness d, width W, length L, and sheet resistance ρ_{sh}.   The series resistance increases linearly with the
sheet resistance and decreases as the square of the number of
grid lines [4], which illustrates the importance of using many
narrow lines rather than a few wide ones.   If horizontal grid
lines are added as well as the vertical ones (Fig. 110b), then
the series resistance decreases as the square of the number
of horizontal grids as well.   Such a checkerboard design could
be useful at high solar intensities where the series resistance
is the major limiting factor.

These same considerations of grid design apply to other
solar cells as well, including CdS, GaAs, $Ga_{1-x}Al_xAs$-GaAs,
Schottky barrier, and heterojunction devices.

Contacts to Si solar cells have been made using electro-
less or electroplated Ni, Au, or Ag, and with evaporated and
sputtered Ni, Au, Ag, Ti, Pd, and Al.   The plating methods have
the greatest convenience and are lowest in cost, but they have
not proven as reliable as the evaporating or sputtering methods.

One of the most widely used contact systems for n-type
Si consists of evaporated Ti stripes followed by evaporated Ag.
The Ti makes a good low resistance Ohmic contact to the Si and
the Ag provides high conductivity along the grid lines.   It
has proven necessary to cover the Ti-Ag stripes with solder
to prevent an electrochemical reaction that takes place between
Ti and Ag in the presence of moisture [265,266], but even with

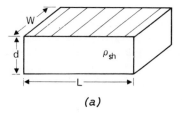 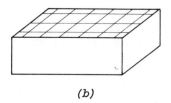

(a)　　　　　　　　　　　　　(b)

FIG. 110. *Series resistances for layers with (a) vertical contact lines, $R_s \propto \rho_{sh}(L/W)/(\# \text{ lines})^2$; (b) both vertical (vl) and horizontal (hl) contact lines, $R_s \propto \rho_{sh}(L/W)/ (\# \text{ vl})^2(\# \text{ hl})^2$.*

the solder some problems can arise because of this tendency for degradation to take place. Recently, a more reliable contact system has been developed having good contact resistance, adherence, and reliability for Si solar cells. This system consists of about 400 Å of Ti to provide low contact resistance and tight adherence to the Si, 5 μm of Ag to provide good electrical conductivity along the stripe, and 200 Å of Pd between the Ti and Ag to prevent the electrochemical reaction [265,266]. This three-layer contact is now widely used for solar cells in the space program.

Aluminum contacts have been made to p-type Si [5,206, 267], and are a particularly good way to contact the p-type bases of N/P cells. The low doping level of 1 and 10 ohm-cm material makes it difficult to obtain low resistance contacts with most metals, but Al dopes the back to a high conductivity and will alloy with it to some degree at normal sintering temperatures (550-600°C), both of which greatly improve the properties of the contact. If the Al is alloyed to the Si at 700-800°C, a BSF contact is obtained with its potential improvements in collection efficiency and open circuit voltage. These BSF's are normally produced by evaporating Al onto the back surface and diffusing for about four hours at 800°C. Back surface field cells have also been made [90] by diffusing boron into the back at 1000°C, but the results were not as satisfactory from all aspects as the Al diffusion method [90].

Pure Ag and Au contacts to the diffused side of the junction have not been used very extensively because of poor adherence [5,206]. Nickel contacts adhere better but have not yielded as low a contact resistance as Ti-Pd-Ag or Al.

Regardless of the contact system used, the contact must be "sintered" at 550-600°C for 5-30 min in an inert atmosphere to prevent high contact resistance and to prevent the possible appearance of a Schottky barrier at low temperatures (-50°C or less) [176,181,183]. Care must be taken not to sinter the upper contact for too long a period or at too high a temperature however because of the possible appearance of a low shunt resistance [23] attributed to the alloying of the metal through the junction and into the base in the vicinity of a microcrack or other defect.

Materials used as Ohmic contacts to GaAs solar cells include Ni, Ag, Al, In, Au-2% Zn, and Ag-2% Zn for the p-side of the device and Ni, Ag, Au-Sn, Ag-Sn, and Au-Ge-Ni for the n-side. The adherence of these metals to GaAs after sintering at 475-550°C for 10 min is better than the adherence of the same metals to Si, but not as good as Ti-Pd-Ag contacts to Si. The Au-2% Zn contact has been widely used as a contact for GaAs lasers, but is difficult to use as the top contact for solar cells with junction depths less than a micron because of a low shunt resistance that can appear after sintering at 475°C or above. The same difficulty of low shunt resistance paths has been noted for In-Zn-Ag contacts [251] sintered at 560°C, while sintering at 450°C instead prevented the shunting problem. Ag-2% Zn forms a reliable, well-adhering contact to pGaAs and can be sintered safely at 500°C for up to 5 min.

For n-type GaAs, Au-Sn or Au-Ge-Ni are most often used. Good results are obtained for Au-Ge-Ni evaporated onto the back of the GaAs wafer and sintered at 450°C and above for several minutes [268]. Au-Sn or Ag-Sn should be sintered at slightly higher temperatures.

Au-2% Zn [9] and Ag-2% Zn have been used as Ohmic contacts to $Ga_{1-x}Al_xAs$ p-type layers also. The contacts are low in resistance after sintering at 500°C for 2-5 min, but the adherence is not as good as it is to GaAs.

Ohmic contacts to Cu_2S-CdS thin film cells are made in a slightly different manner than other types of solar cells. A foil of Mo or a plastic such as Kapton is coated with Ag, Ag-Zn [77], Ti-Pd-Ag [54], or another thin metal film and a 1-mil CdS film is evaporated onto this metal. After forming the Cu_2S layer on the CdS surface by dipping in CuCl solution, the structure is heat treated, which produces a reasonably low resistance contact between the CdS and the underlying metal. Copper or gold grids are then evaporated or electroplated onto the Cu_2S surface or physically laid onto this surface and held there by the subsequent lamination. The conductivity of the Cu_2S is so high that low resistance Ohmic contacts are automatically produced in this way.

At high solar concentrations, the voltage drop along the metal grid line can become a major problem due to the high current outputs. Low resistivity grid lines made with Ag, Cu, Au, or Al will probably be needed, and the grid thickness and width will have to be increased.

H. Organic Solar Cells

Solar cells made from organic materials have received little attention in the past, mainly due to their low measured efficiencies (0.1% or less). The ease of fabrication of solar cells made from these materials together with their possible low cost warrants a closer look at their potential as useful devices. Organic solar cells have been made from anthracene [269], tetracene [270], phthalocyanine [271-274], and chlorophyll [275,276]. In all these devices, the power conversion efficiency has been limited by low quantum efficiencies (poor collection of photogenerated carriers) and not by low voltage output or fill factors. The poor quantum efficiencies in turn are probably due to a very high trap density, which lowers the lifetime, mobility, and diffusion length to poor values. Other difficulties with organic solar cells include the high resistivities of these materials (10^5 ohm-cm or higher) and the problems of making Ohmic contacts to them.

Organic solar cells are usually prepared by vacuum evaporation. A metallized glass or metallic substrate which has been carefully cleaned is placed in a vacuum of 10^{-5}-10^{-8} Torr and a film of organic material is evaporated onto it from a source held at 150-200°C. The substrate is usually unheated, and the deposited organic film is from 1000 Å to 1 μm thick. A transparent metal or conducting glass electrode is then deposited onto the surface of the film, either with or without exposing the film to the atmosphere first. Each step along the way is capable of having a strong effect on the resulting device, and each must be carefully controlled to obtain reproducible results.

The optical absorption coefficients of tetracene [270], Mg-phthalocyanine [274], and chlorophyll [276] are shown in Fig. 111. Materials such as phthalocyanine have broad absorption in the 5000-9000 Å range of wavelengths, with absorption coefficients of 10^5 cm^{-1} over much of the visible region; films of about 1000 Å will absorb most of the sunlight in this range. Tetracene has a rather narrow absorption band centered around 5000 Å. Chlorophyll has strong absorption around 4500 and 7400 Å, with weaker absorption in between; films of 1000-2000 Å of this material will absorb 20-30% of the overall sunlight. Since the diffusion lengths in organic materials are quite low, the materials with the highest absorption coefficients over the visible region are most likely to have the highest quantum efficiencies and photocurrents.

Most organic solar cells have been of the Schottky barrier type using a transparent metal such as Al to form the barrier (Fig. 112). Light enters the organic material through the

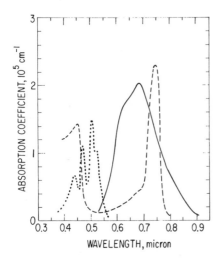

FIG. 111. Optical absorption
coefficients of tetracene (···),
Mg-phthalocyanine (——), and
crystalline chlorophyll (---).

metal film and is absorbed as a function of distance in accordance with the absorption coefficient. Electron-hole pairs
may be created directly [274], but it seems more likely that
excitons (loosely bound hole-electron pairs) are created first
and are then separated into individual carriers either at
impurity centers [272,274] or in the high field barrier region
[269-271,277]. Either way, photocarriers will be collected
over a distance equal to the barrier space charge region width
plus about 1 diffusion length from the edge of this region
(the effective diffusion length would be larger if an aiding
drift field were present in the organic film). Since the space
charge region width is around 200 Å and the diffusion lengths
are about the same [274], only the light absorbed in the first
400-500 Å will contribute to the photocurrent.
      Ghosh et al. [270,274], Tang and Albrecht [275,276], and
others have compared the spectral response obtained when light
is incident on the barrier side with the response obtained
when light is incident on the Ohmic contact side (Fig. 113).
For front illumination, the more strongly absorbed the light
is, the more it will create carriers within this W+L region
and the higher the spectral response will be. For back illumination, the more strongly absorbed the light is, the farther
it will generate carriers from the W+L region, and only weakly
absorbed light can contribute efficiently to the photocurrent.
The spectral response is then in some ways related to the
reciprocal of the absorption coefficient, with the peaks in
the response occurring at the wavelengths where the most carriers are generated in this region. The thinner the organic
film is relative to W+L, the more the responses obtained from

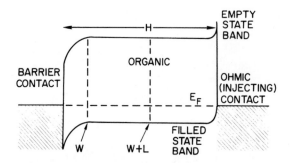

FIG. 112. *Energy "band" schematic of metal (Schottky barrier)-organic film-metal (Ohmic contact) solar cell. H is the width of the organic film, W is the space charge layer width, and L is the diffusion length.*

the two sides will be alike. This fact can be used to estimate the diffusion length, using the zero bias capacitance to obtain the space charge layer width.

The short circuit photocurrents under white light illumination for devices made from undoped, high resistivity organic materials have ranged from 10^{-5} to 10^{-4} mA/cm^2 for about 10^{-1} mW/cm^2 input intensity [270,273,274]. Unlike semiconductor (inorganic) solar cells, the photocurrent in organic devices does not increase proportionally with the input intensity F, but increases instead as a power of the intensity F^n where n is around 1 at low intensities and decreases to 0.5 as the intensity increases [270,272,274,275]. This power law dependence has been attributed to the presence of a high density of traps in the organic film [270,274]; the statistics associated with trapping and recombination processes depend strongly on the density of photoexcited carriers in high resistivity materials when a high trap density is present.

It appears that the photosensitivity of Mg-phthalocyanine films is strongly enhanced by "doping" the films with oxygen, and possibly by doping with Al also [273]. Usov *et al.* [271-273] have studied Al/Mg-Ph/Ag devices prepared under various conditions. When such devices were first made, without ever exposing them to oxygen, the resistivity of the organic film was very high (10^{11} ohm-cm) and the photosensitivity was low, reportedly about 3×10^{-5} mA/cm^2 for 0.5 mW/cm^2 white light illumination. After exposing the devices to air for 5 min, and heating to 50°C in vacuum, the resistivity decreased to

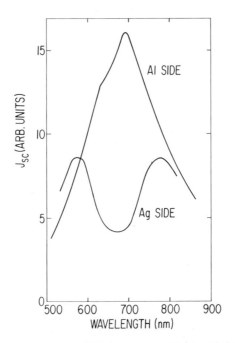

FIG. 113. *Spectral responses of an Al/Mg-Ph/Ag cell for light incident on the barrier (Al) side and for light incident on the back contact (Ag) side, H = 1500 Å. (After Ghosh et al. [274]; courtesy of the American Institute of Physics.)*

10^8 ohm-cm and the photocurrent had increased by two orders-of-magnitude to 3×10^{-3} mA/cm^2 at the same intensity. The rectification ratio was increased from less than 2 to about 40 by this oxygen treatment. As a third preparation step, the devices were then heated to 80°C for 1-2 hr in vacuum with a 0.5 V bias applied; the photosensitivity remained the same but the rectification ratio increased to over 1000. The explanation given was that the Al may have diffused into the organic film and formed a p-n junction with the normally p-type Mg-Ph [273].

The peak quantum efficiency of the oxygen-treated Mg-Ph devices, after the second step, was reported to be as high as 0.3-0.4 at 6900 Å, much higher than the 0.001-0.002 quantum efficiencies obtained with most organic materials [271,272].

The electrical properties of organic solar cells in the dark have been studied with both current-voltage and capacitance voltage techniques. At low voltages (<0.4 V), the current

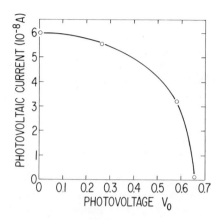

FIG. 114. *Current-voltage behavior of an illuminated Al/ tetracene/Au solar cell for 0.1 mW/cm^2 intensity. (After Ghosh and Feng [270]; courtesy of the American Institute of Physics.)*

obeys the familiar exp(qV/AkT) relationship with A ranging from 2 to 3 [272,273]. At higher voltages, the current varies as a power of the voltage, $J \propto V^n$, with n ranging from 2 to 4 [270,272,273,275]. This type of current-voltage behavior is strongly indicative of traps [270,278], and trap densities in the 10^{18}-10^{19} cm^{-3} range are implied by the data. The dominance of organic devices by traps is also indicated by the frequency dispersion of the capacitance; the zero bias capacitance of Mg-Ph units, for example, decreased by a factor of 10 from 0.1 to 10^4 Hz [272], which can be explained [278] by the increasing inability of deep traps to charge and discharge in time with the ac signal as the frequency increases. Other phenomena which imply trapping processes include the slow decay of the photocapacitance [274] and the slow response (rise and decay) of the photocurrent to incident light pulses [271,274] (after an initial fast rise or decay at higher amplitudes).

The current-voltage characteristics of an illuminated tetracene solar cell [270] are shown in Fig. 114. The open circuit voltages of organic cells have ranged from 0.35 V for chlorophyll [275] to 0.85 V for phthalocyanine [274]. The FF at low light intensities can be quite good (up to 0.75) as long as the resistivity of the organic material is not too high. The series resistance contributed by a 1000 Å thick layer of 10^8 ohm-cm material, for example, is only 1000 ohm for a 1 cm^2 area. Since the internal impedance of the current generator (Fig. 31) is V_{mp}/J_{mp} and is about 10^6-10^7 ohm for

the same area, the device is not significantly affected by
series resistance at these low intensities. If the resistivity
of the material is much higher, however, or if the intensity
is increased to 10-20 mW/cm^2 with a good quantum efficiency,
the series resistance will be more important and the FF will
be reduced accordingly.

In addition to the bulk resistivity, a serious contribution to the series resistance could come from the back, supposedly Ohmic, contact. It is notoriously difficult to make
a low resistance contact to high resistivity materials. In
Fig. 112, an "injecting" contact to a p-type organic film is
shown; this is the best possible case for an Ohmic contact to
these films, similar in nature to a BSF contact (Fig. 7). In
the worst case, a Schottky barrier might be present at the
back, which would lead to a very high, voltage-dependent series
resistance, and which would also lower the FF, reduce the open
circuit voltage, and lower the short circuit current.

The power conversion efficiencies of organic solar cells
have been quite low. For Al/tetracene/Au cells, Ghosh and
Feng [270] report an efficiency of 10^{-4}% for "white light" of
low intensity, and Ghosh et al. [274] report white light efficiencies of around 10^{-3}% for Al/Mg-Ph/Ag cells with the undoped,
un-heat-treated form of phthalocyanine. Doped Ph films are
more photosensitive, and Federov and Benderskii [273] have
given data which lead to an efficiency of 0.1-0.2% for white
light illumination of 0.5 mW/cm^2. This implies an average
quantum efficiency of around 0.01 over the visible region,
with a peak quantum efficiency of 0.1-0.2.

The efficiencies of the organic solar cells made so far
have decreased with increasing intensity due to the $J \propto F^n$
power law relationship of the photocurrent to the input intensity, and possibly due to a decreasing FF as well. The key
to raising the conversion efficiencies lies in raising the
quantum efficiency, which in turn depends on reducing the trap
density by better purification of the material. The quantum
efficiency can also be improved by using materials with the
highest absorption coefficients, and trapping effects can be
reduced to some extent by doping the material to obtain lower
resistivities. A lower bulk resistivity would also facilitate
making an Ohmic contact to the back of the cell. If the quantum efficiency can be brought to an average of 0.1 or higher
over the visible spectrum, efficiencies of several percent at
AM1 should be readily attainable [279]. Organic solar cells
made from chlorophyll are particularly intriguing, since this
is one of the most common substances found on earth.

TABLE 23
Abundance of Certain Elements in Earth's Crust
(Not Including Ocean or Ocean Bottom)[a]

Element	Abundance (ppm)	Element	Abundance (ppm)
Si	276,000	Pb	12.0
Al	80,000	As	2.0
Fe	50,000	Sn	1.7
S	1,500	W	1.0
P	960	Cd	0.18
Cu	64	In	0.14
Ga	18	Se	0.09

[a]After Ref. [286].

I. Abundance of Materials

In order for photovoltaic solar energy conversion to
make a significant impact on the energy needs of the United
States or any other part of the world, there are at least
three important conditions it must satisfy. First, the cost
of generating energy using photovoltaics must be economically
competitive with other available means of producing energy,
although some allowance can be made for the safety and environmental compatability benefits obtained with solar energy.
Second, the amount of energy obtained during the life of a
photovoltaic system must be much larger than the energy needed
to fabricate and operate the system. Little attention has
been given to this point in the past, and present-day methods
of fabricating solar cell arrays are very lossy. Third, there
must be enough material available to fabricate solar cell
arrays in sufficient quantity to generate a significant fraction (greater than several percent) of the country's energy
needs.

The potential cost of solar cell arrays has been discussed
very thoroughly recently [190,280-284], and the energy balance
problem has also been described [280,283-285]. The issue of
material availability will be discussed here.

The electrical energy demand in the United States amounts
to about 2×10^{12} kW-hr/yr as of 1975 [284], and can double by
1990. This represents about 3×10^{11} W of required generating
capacity on the average, and two to three times this capacity
must be available to accommodate peak demand. A practical
photovoltaic system must be able to supply a significant

TABLE 24
Mineral Production in Metric Tons = 10^3 kgm[a]

Mineral	US Prod/yr	Cost ($/kg)	World Prod/yr
Al	2×10^6 (1971)	0.53	5.6×10^7 (1971)
As	2.9×10^3 (1968)[b]	1.21 (pure)	5.2×10^4
Cd	3.0×10^3 (1970)	7.00	1.7×10^4 (1970)
Cu	1.4×10^6 (1970)	1.32	4.4×10^6 (1970)
Fe	1.2×10^8 (1971)	0.22	---
Ga	0.3 (1968)	750.00	1.0 (1968)
In	7.15 (1967)	80.00	62.2 (1968)
P	10^7 (1970)	1.32	---
S	9.5×10^6	0.033	---
Si[c]	1.25×10^3 (1974)	35–60	---
Se	3.4×10^2	30.00	5×10^2
W	4×10^5 (1970)	9.91	---
Zn	5×10^5 (1970)	0.33	5×10^6 (1970)

[a]From U.S. Dept. of Interior [287], and NSF-RANN report NSF-RA-N-74-072 [288].
[b]1.6×10^4 additional were imported.
[c]Semiconductor grade.

fraction of this, say 10^{10} W under peak conditions, and perhaps five times this amount should be a reasonable goal to take into account the losses due to poor weather, nighttime, and the need for energy storage. The two chief contenders to meet this requirement are ribbon Si and thin film CdS, with thin film Si and thin film GaAs as possibilities and other materials such as InP, $CuInSe_2$, and even organics as "dark horses."

The average crustal abundances in parts per million of various elements in the earth's crust are shown in Table 23. Silicon, Al, and Fe are very abundant, while Cd, In, and Se are relatively rare, with Ga, Pb, As, and Sn in between. Since there are 2.5×10^{16} metric tons in the first kilometer of the continental United States crust alone, there is theoretically enough of any material to generate 10^{16} W or more. However, it is the supply of materials and the amount that can be economically obtained that are important, not the crustal abundance.

The mineral production in the United States and in the world for the last several years is shown in Table 24, together with the cost per kilogram in the United States. The fabrication of CdS, GaAs, and InP devices is limited by the availability of Cd, Ga, and In, respectively. The relatively large quantity of semiconductor grade Si produced each year is used

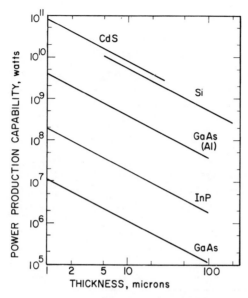

FIG. 115. *Peak power production capability of several materials using present production rate in the United States. 10% efficiency, 100 mW/cm^2 input power. No concentration.*

almost entirely by the semiconductor industry, and this amount will have to be increased considerably to supply solar cell needs. The power generating capacity that could be obtained if the yearly production of Cd, Ga, In, and Si were converted entirely into solar cells of various thicknesses is shown in Fig. 115. The present level of Ga production could only yield 10^6 to 10^7 W under peak conditions, with InP about an order-of-magnitude higher. Ribbon Si 4 mil thick could yield around 5×10^8 W, while thin film CdS or Si devices 10 μm thick could yield about 10^{10} W.

The power generation capability of GaAs devices using the Ga contained in the yearly production of Al is also shown in Fig. 115. Aluminum ores contain about 50 ppm of Ga on the average; almost all of this is thrown away as a waste product. If the Ga were recovered instead, over 10^9 W could be produced with 1-2 μm thick devices. Such a large new market for Ga would probably drop its price considerably, perhaps to as low as that of In.

If the present yearly production rate of these minerals is maintained, only CdS and thin film Si are capable of generating the minimum requirement of 10^{10} W in a relatively few years. Ribbon Si and thin film GaAs would take about 10 yr at the present rates to equal this generating capacity. Either the present production rate of the relevant minerals will have

TABLE 25
Identified Mineral Resources[a]

| Mineral | U.S. Resources | | World Resources | |
	Ref.[288]	Ref.[287]	Ref.[288]	Ref.[287]
Al[b]	---	2×10^{11}	---	3.5×10^{12}
As	1.4×10^{6}	1.1×10^{6}	1.9×10^{7}	1.8×10^{7}
Cd	2×10^{5}	3×10^{5}	8.5×10^{5}	1.2×10^{6}
Cu	9.1×10^{7}	6.5×10^{7}	4×10^{8}	2.9×10^{8}
Fe	---	1.2×10^{11}	---	2×10^{12}
Ga	2.7×10^{3}	8.4×10^{3}	1.1×10^{5}	1.1×10^{5}
In	5×10^{2}	5.8×10^{3}	3.2×10^{3}	9×10^{3}
P	---	2.9×10^{9}	---	5.1×10^{10}
S	---	2.9×10^{10}	---	---
Se	2×10^{4}	1.4×10^{5}	1×10^{5}	2.5×10^{6}
Si	Unlimited		Unlimited	
W	---	2.9×10^{6}	---	5.1×10^{7}
Zn	---	1.2×10^{8}	---	1.5×10^{9}

[a]Potentially economically and technologically recoverable at today's market (in metric tons).
[b]Includes all Al ores.

to be increased severalfold (to supply the present use of these materials as well as solar cell needs), or a rather slow build-up of photovoltaic power generation capacity will have to be tolerated.

The identified resources of various minerals in the United States and in the world have been estimated by the Department of the Interior, and are given in Table 25. These are conservative estimates for the available resources; estimates which include unidentified sources run much higher [287]. The power production capability of various materials that would be obtained if all the identified resources of the limiting minerals were converted into solar cells is shown in Figs. 116 and 117. The Si availability is nearly unlimited, and the generation capability of Si is off scale in these figures. The generation capability of CdS is well above the 10^{10} minimum, and the capability of GaAs using the conservative resource estimate is also above this minimum. If the Ga content of the estimated Al resources in the United States is used, the GaAs capability is much higher (Fig. 117, dotted line). InP is also capable of meeting the minimum requirement. There may be rich new sources of In and Ga in coal burning residues and from coal gasification, which could make the prospects for InP and GaAs look even better.

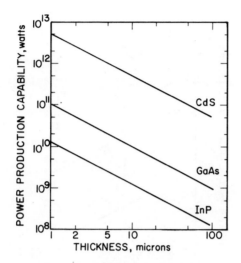

FIG. 116. Peak power production capability of several materials using identified United States resource estimates. Silicon is off scale. For GaAs from Al resources, see Fig. 117. 10% efficiency, 100 mW/cm^2 input power. No concentration.

The calculations of Figs. 115-117 have been made for 1 incident solar intensity. It should be economically and technically feasible to concentrate sunlight onto solar cells by at least a factor of 20 using reflectors made from cheap materials such as Al. The cost of the array would then be decreased while the generation capability of a given area of solar cells would be increased. GaAs is particularly attractive from this point of view, since good efficiencies can be obtained at high temperatures and intensities [289]. Silicon and InP could also be used at relatively high intensities. CdS may not be as attractive in this regard because of the adverse effects of temperature.

The land area taken up by a photovoltaic solar energy system is about 10^7 m^2 (10 km^2) for every 10^9 W capacity at 10% efficiency and 100 mW/cm^2 input intensity. To meet the 10^{10} W minimum requirement would take about 100 km^2, and if the losses due to poor weather and nighttime are allowed for, a 10^{10} av W capacity would take about 500 km^2. The entire present electrical consumption of the United States of 3×10^{11} W could be generated by a photovoltaic system of 3000 km^2 at 10% efficiency and 100 mW/cm^2 input, or 15,000 km^2 allowing for losses. This last figure represents 0.19% of the land area of the United States, and represents 30% of the land area now covered by roads [283].

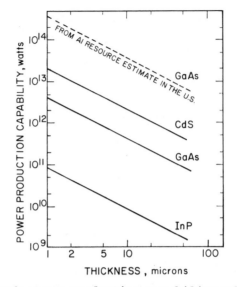

FIG. 117.   *Peak power production capability of several mate-*
rials using identified world resource estimates. Silicon is
off scale.   10% efficiency, 100 mW/cm^2 input power. No con-
centration.

J.   Summary

In this chapter, an introduction has been given into the
various technologies used in the fabrication of solar cells.
Crystal growth methods for Si substrates include the Czochralski, FZ, and ribbon-growth techniques, while for GaAs substrates the horizontal Bridgman method is most often used.
All of these techniques involve growth by controlled freezing
at a solid-liquid interface. Dopants can be added directly
to the molten Si or GaAs, and will be incorporated into the
growing solid in accordance with the segregation coefficient.
Boron and phosphorus are the dopants most often used for Si,
and Se, Te, or Si are most often used for n-type GaAs.

Chemical vapor growth is receiving increasing importance
in solar cell technology. Vapor-grown layers can be used as
an alternative to diffusion as a means of forming the thin
top region of the cell, or they can be used in fabricating
thin-film single crystal and polycrystal solar cells for wide-
scale terrestrial use. Dopants are easily incorporated from
the vapor phase.

The diffusion step is one of the most critical ones for
conventional cell fabrication. Very shallow junctions with
high surface concentrations are needed for high efficiencies.

The stresses, defects, and electrically inactive impurities often introduced by the diffusion can result in a "dead layer" of very low lifetime near the device surface. To eliminate the dead layer, slightly lower surface concentrations and even shallower junctions are helpful, and the use of dopants with better matches to the host lattice would also be beneficial. Conventional N/P silicon cells are usually made by diffusing phosphorus or As into boron-doped substrates. Most GaAs cells are made by diffusing Zn into n-type GaAs substrates.

Thin film CdS solar cells suffer from a number of degradation problems, including attack of the Cu_2S by water vapor or air, a light-activated phase change of the Cu_2S at high temperatures, and an electrochemical breakdown of the Cu_2S with the formation of metallic Cu if the load voltage exceeds 0.35 V. Considerable progress has been made in understanding and minimizing these problems, however, and useful device lives of 20 yr or more have been predicted under carefully controlled conditions.

Heterojunctions are usually made by the vapor growth or LPE of a wide gap material on a Si or GaAs substrate. Good lattice match, good thermal expansion match, and the absence of significant undesirable cross doping effects are prerequisites for obtaining good heterojunction cells. Schottky barrier cells are easier to fabricate, provided the metal films are evaporated under high vacuum. Metal layers 50-100 Å thick with AR coatings have good sheet resistivities and good (>90%) optical transparency. Transparent conducting materials such as In_2O_3 could be useful for both heterojunction and Schottky barrier cells.

Ion implantation as a solar cell fabrication technique has the advantage that shallow junctions with high surface concentrations can easily be obtained, and the doping profile in the top region can be tailored to produce an aiding drift field. The disadvantage is that the lattice damage produced by the implantation is difficult to anneal, and low open circuit voltages are therefore usually obtained.

Antireflective coatings are important in reducing the large fraction of the incident light reflected from a bare Si or GaAs surface. Single layer AR coatings can reduce the average reflection from about 40% (bare material) to around 10%, and double layer coatings can further reduce the reflection to around 3%. For single layer coatings, a material should be used which has low optical absorption and which has an index of refraction equal to or slightly larger than the square root of the refractive index of the semiconductor substrate.

A wide variety of metals and alloys can be used as Ohmic
contacts to Si and GaAs, with high electrical conductivity
along the stripes, low contact resistance to the semiconductor,
and tight adherence to the semiconductor as the important
criteria. Contact sintering is necessary to produce low contact resistance but must be done carefully to prevent shunt
resistance problems. The contact grid pattern should be designed carefully; the series resistance decreases as the square
of the number of grid lines. The conductance of the metal
grid itself becomes a significant problem at high intensities.

Organic solar cells have been low in conversion efficiency
because of poor carrier collection. The key to improving these
devices lies in reducing the trap density, lowering the resistivity, and providing good Ohmic contacts.

ADDENDUM

Recent Results

During the early summer of 1975, while the main portion
of this book was being printed, several important conferences
in solar energy took place, and a number of new and interesting results were presented. The most important of these sym-
posiums from the photovoltaics point of view was the 11th IEEE
Photovoltaics Specialists Conference held in Scottsdale in
May. A Workshop on CdS solar cells was held in Delaware, and
a symposium on materials and processes for solar energy conversion was held as part of the Electrochemical Society spring
meeting in Toronto. Other conferences sponsored by the American Vacuum Society were held in Yorktown Heights and Boston.
Most of these conferences will publish proceedings or extended
abstracts, and a few of the important details presented at
these meetings will be discussed briefly here.

Some of the most important recent results were centered
around the new Si solar cell produced with a serrated surface
(Chapter 10, Section F); this cell is now designated as the
CNR cell for Comsat NonReflecting. The main features of this
device are shown in Figs. 118 and 119. The concept is a rather
radical departure from the very flat, polished surfaces used
in the past. The pyramidal surfaces can be produced on <100>
oriented wafers by preferential etching techniques (a hydra-
zine-hydrate etch for 10-30 min was discussed by NASA-Lewis).
The benefits obtained are reduced reflectivity, less dependence on the antireflection coating, and increased collection
efficiency due to carrier generation closer to the junction.
Figure 119 illustrates these points in one dimension. Light
incident on the side of a pyramid can be reflected onto another
pyramid instead of being lost, provided that the pyramid base
angle is around 45° or more. The reflectivity of bare Si is
reduced from 35 to 45% for flat surfaces to around 20% for the
serrated surface, and the addition of an antireflection coating reduces the overall reflection to a few percent.

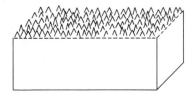

FIG. 118. Serrated surface of a nonreflecting CNR cell.

The increased collection efficiency is a result of the refraction of the light entering the Si. Carriers are generated on the average closer to a collecting junction boundary, which makes the cell less dependent on the base diffusion length than a planar cell and gives it high radiation tolerance. Some degree of total internal reflection may take place at the back of the cell, which improves the long wavelength response.

CNR cells with very narrow junctions (as in the "violet cell") and with pyramidal heights and spacings of 10 μm and 5-10 μm, respectively, have yielded photocurrents of 45 mA/cm^2 (180 mA for a 4 cm^2 device) compared to 38-40 mA/cm^2 for planar cells. The dark I-V characteristics are reported to be excellent, and AM0 efficiencies of over 15% have been obtained so far. Work on these cells is being done at Comsat Laboratories and NASA-Lewis and will probably be started in many other laboratories shortly.

High doping level effects in Si solar cells was another subject of considerable interest. Conventional theories predict that the voltage output and power conversion efficiency will *increase* with decreasing base resistivity down to about 0.1 ohm-cm; they also predict that the doping level in the diffused region will have little effect on the voltage output, since the dark current contributed by the base region of an abrupt junction device will be much larger than the dark current contributed by the top region. Experimentally, all this seems to be true down to about 1 ohm-cm bases for violet cells, but unexplained behavior occurs for more heavily doped bases and for heavily doped diffused regions; the open circuit voltage and efficiency saturate or even decrease with base resistivities of less than 1 ohm-cm. Several papers were presented in which this was attributed to a combination of bandgap shrinkage [290], extremely low lifetimes in the diffused region, and the linearly graded nature of the junction.

The bandgap shrinkage is due to the very high surface concentration of impurities in the diffused region, 5×10^{20}–10^{21} atoms/cm^3 in nonviolet cells, which can result in an energy band configuration (electric field) which forces carriers toward the surface rather than toward the junction. The smaller bandgap would also cause higher absorption and

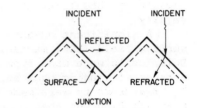

FIG. 119. *Reflection and refraction of light incident on a serrated surface.*

carrier generation near the surface, where surface recombination and poor lifetimes would reduce their collection probability.

The short lifetime in the diffused region is the familiar "dead layer" problem that has been largely solved by the low doping level, narrow junction principles of the violet cell, but which is present in conventional cells. For high doping levels the lifetime may decrease by as much as N_d^{-2} due to the strain, dislocations, point defects, and unwanted impurities introduced by the diffusion process. Very low lifetimes would have the effect of lowering the photocurrent and lowering the open circuit voltage by increasing the dark current.

The linearly graded junction model is also important in explaining the effects of high doping on the dark current. In an abrupt junction the current due to minority carriers injected from the base into the diffused region would be negligible, but in the graded junction the current injected from the base would be comparable to and perhaps even larger than the current injected into the base. Increasing the *base* doping level, according to the theory, increases this dark current due to injection from the base into the diffused region, lowering the open circuit voltage. Increasing the doping level in the *diffused* region also increases the dark current because of the bandgap shrinkage and the very low lifetimes (100 psec), and again the open circuit voltage is reduced.

There are also other possible explanations for high doping level effects on voltage outputs from solar cells including tunneling, excess space charge recombination, Auger recombination, and possibly field and mobility gradient effects. Which of these is most important is still open to question at this point. Experimentally, the open circuit voltage in N/P Si cells saturates at 0.625-0.635 V for around 0.1 ohm-cm base material, instead of reaching the 0.70 V predicted by conventional theories (Table 6). The efficiency peaks at a base resistivity between 0.1 and 1 ohm-cm for conventionally diffused cells, but probably closer to 0.1 ohm-cm in violet cells.

Contributions in the understanding of high doping effects were made by the University of Florida, the University of

Illinois, North Carolina State University, Optical Coating
Laboratory, AEG-Telefunken, NASA-Lewis, and Wayne State University.

A significant advance in the understanding of the back
surface field (BSF) mechanism has been made at NASA-Lewis.
Solar cells were made from 1, 10, and 100 ohm-cm p-type substrates, including back surface fields. Conventional V_{oc} and
I_{sc} measurements were made under AM0 conditions. The front
n^+-p junction was then etched off, leaving only the p-p$^+$ junction at the back. Ohmic contact was made to the front, and
solar cell measurements were made once again. A 100 mV V_{oc}
was obtained with 100 ohm-cm material, and a 10 mV V_{oc} was
obtained with 10 ohm-cm material, establishing that some of
the increased output (but not all) obtained from BSF cells is
due to voltage generation at the high-low junction. In a
10 ohm-cm cell, the open circuit voltage is increased from
0.55 V to 0.6 V; 0.01 V is therefore due to a voltage generation at the back, and 0.04 V is due to the benefits of the
carrier confinement in reducing the overall dark current.
Cells made from 100 ohm-cm material with a BSF had the same
0.6 V output as 10 ohm-cm BSF cells. Cells with 1 ohm-cm bases
apparently have no appreciable voltage generated at the high-
low junction, but may still benefit somewhat from confinement
provided the base diffusion length exceeds the cell thickness.

Advances in Si materials technology and in nonconventional
Si devices were reported also. Solar cells made from EFG
ribbon were reported by Mobil-Tyco, and cells made from dendritic web were described by Westinghouse. Efficiencies of
10% at AM1 have been achieved with the ribbon and 11-12% with
the web Si. One of the problems here seems to be that arrays
of parallel slip bands or twins or other line defects are
sometimes present in these materials and act as zones of high
recombination, decreasing the short circuit current. When
such parallel line defects are absent, considerably better
short circuit currents are obtained [290a].

Experiments in processing metallurgical-grade Si into
solar cells were described by Dow Corning. Czochralski ingots
were pulled from relatively crude metallurgical Si rather than
high purity semiconductor grade, relying on the low segregation coefficients of most elements in Si to self-purify the
Si ingot as it was grown. "First" ingots had impurity contents
around 100 times less than the starting material. When selected
areas of these were remelted and a second ingot grown, higher
purities were obtained and the material could be used to fabricate reasonably good solar cells. In another technique, chlorine
and oxygen were passed over the surface of the molten Si, and
the impurity content of the melt was reduced by the formation of

volatile chlorides, oxides, and other compounds. The resulting
Si could then be gradient frozen to obtain further purification
by segregation.

Thin film solar cells made on carbon and metallurgical
grade Si substrates were described by Southern Methodist University. It was discovered that "recrystallization" [291],
annealing of the grown Si layer near or even above the melting
point, could result in much larger grain sizes (100 μm) than
obtained in the initial Si layer (5-10 μm). AMO efficiencies
of 2.5% were obtained from cells made from polycrystalline
layers grown on metallurgical Si substrates. The efficiency
was probably limited by remaining small grains within the
large grains.

A review of potentially low-cost processing methods for
Si photovoltaics was presented by M. Wolf of the University
of Pennsylvania. Such methods include the ribbon and dendritic
web crystal techniques, hot rolling and extrusion to form Si
sheets, continuous Si production from sand and carbon using
SiF_2 transport, sheet casting, and even solution growth. Many
of these fast sheet production methods will likely result in
polycrystalline material, so that recrystallization techniques
to obtain sufficiently large grains are important. There is
much room for interesting materials work here; past processes
have optimized material quality at the expense of cost, energy
used, and speed, while future processes will have to optimize
speed per unit cost per unit energy for a sufficiently pure
solar grade quality capable of making a 10% efficient cell.
Continuous, flowthrough processing for both materials preparation and device fabrication will have to be emphasized if
low-cost cells are to be made. Devices used in concentrated
sunlight schemes, however, could still be produced by efficient batch processing techniques.

With well-designed, continuous processing, the energy
payback time (the time needed for a device to generate as much
energy as used to fabricate it) will be about 1/2 year. With
concentrated sunlight schemes, the payback time could be considerably less.

Recent results on radiation damage to solar cells in the
space environment were presented, including neutron, electron,
and UV irradiation tests. The CNR cells were reported to be
more resistant to radiation degradation than the conventional
planar violet cell, which is already a radiation tolerant
device. Vertical multijunction (VMJ) cells were described
which consist of separate parallel vertical slabs (not filled
in between as in Fig. 81c), combining the CNR nonreflecting
surface concept with the inherent high radiation tolerance
of VMJ devices. A single paper was presented on Li-doped Si

cells. There seems to be a de-emphasis lately on Li-doped
devices; the speculation is that these devices are only slightly
better than violet cells and cannot justify their added cost
and fabrication complexity.

In the "New Approaches" session results on Schottky barrier cells, high-efficiency GaAs cells, induced junction devices, heterojunctions, and p-i-n structures were reported.
In Schottky cells the evidence continues to grow that a thin
interfacial layer can improve the performance significantly.
Several theoretical pictures were presented, including one
that attributes the higher output voltages to the presence of
surface states, fixed charge, and traps within the interfacial
layer in a certain configuration. Data were presented by
Stirn of JPL on Au-GaAs Schottky barrier cells; improvements
of V_{OC} from 0.45 to 0.70 V could be achieved by heat treating
the GaAs substrate prior to depositing the semitransparent
metal layer. Efficiencies of 15% in terrestrial sunlight were
seen in some small area cells. Apparently this increase in
V_{OC} can be obtained without adversely affecting the short circuit current or fill factor, at least for oxides of the order
of 50 Å thick or less. Similar improvements in device behavior
by interfacial oxides have been seen in Si devices as well
[292,293]. Contributions in the Schottky cell area came from
the University of British Columbia, Pennsylvania State University, the Joint Center for Graduate Study, and the Jet Propulsion Laboratory.

Heterojunction and Schottky barrier solar cells made with
In_2O_3 or SnO_2 on n- and p-type Si were described by a joint
Innotech Corporation-Syracuse University group. Efficiencies
at AM1 of 6% were seen on In_2O_3-pSi devices, with open circuit
voltages of 0.30 V and short circuit currents of 35 mA/cm^2.
The electron affinity of In_2O_3 was estimated to be about 4.34 eV;
this limits the V_{OC}'s and η's that can be obtained from In_2O_3
heterojunctions on p-type Si to about 0.4 V and 11%, respectively, and effectively rules out the possibility of good
Schottky barrier cells (nIn_2O_3-nSi heterojunctions) due to a
maximum V_{OC} of about 0.15 V. The electron affinity of SnO_2,
however, is much higher (4.75 eV), and therefore, good Schottky
barrier cells ($nSnO_2$-nSi) are possible, but good heterojunctions ($nSnO_2$-pSi) are probably not. V_{OC}'s of up to 0.35 V and
J_{SC}'s of up to 25 mA/cm^2 were seen on $nSnO_2$-nSi devices (about
1 ohm-cm Si resistivity). The fill factors were lower in these
cells (0.4) than in In_2O_3-pSi cells (0.6) for reasons which
are not clear at this time.

The transparent conductor fabrication techniques have
been applied successfully to a variety of Si substrates. In_2O_3
films on p-type Si have resulted in: $V_{OC} = 0.33$ V, $J_{SC} =$

30 mA/cm^2 for 1.4 ohm-cm Czochralski wafers; V_{OC} = 0.30 V, J_{SC} = 36 mA/cm^2 on 9 ohm-cm Tyco ribbon; V_{OC} = 0.31 V, J_{SC} = 30 mA/cm^2 on 1 ohm-cm polycrystalline wafers (large grain sizes); and V_{OC} = 0.085 V, J_{SC} = 10 mA/cm^2 on 0.05 ohm-cm metallurgical grade Si wafers. All these were at AM1 conditions, 100 mW/cm^2 input.

GaAs solar cells received some attention also. J. Ewan of Hughes Research Laboratories described a liquid-phase epitaxy system for making large numbers (30-40) of large area ($\gtrsim$2 cm^2) Ga$_{1-x}$Al$_x$As-GaAs cells on a semicontinuous basis, while J. Hutchby of NASA Langley and K. Takahashi of Tokyo Institute of Technology described theoretical predictions for a graded bandgap Ga$_{1-x}$Al$_x$As-GaAs device. An IBM Research group described a technique for making good GaAs cells (18% AM0) out of poor quality substrates by a combination of leaching in Ga to remove recombination centers and fabricating cells with deep (1-1.5 μm) instead of shallow (0.3-0.5 μm) junctions. Measured efficiencies of 14.7% at AM0 and 19% at AM1 have been obtained in this way. The devices also behave well at high temperatures; the measured AM0 efficiency drops to 9% at 250°C and 6% at 300°C.

A group from Varian Associates described their results on high sunlight concentrations on Ga$_{1-x}$Al$_x$As-GaAs cells [294]. Efficiencies of 19% have been obtained for devices operating at 1735 solar intensities, and 23% efficiencies have been measured at AM1.4 at low concentrations (10). The output at 1735x amounted to 240 kW/m^2 of cell area. Operation of Ga$_{1-x}$Al$_x$As-GaAs cells at up to 5000 suns has been reported [295]. The importance of this work lies in its terrestrial implications. Optical concentrators are much less expensive than solar cells, and calculations indicate that solar power generation can be made economically viable even using the highly expensive, high-performance GaAs solar cells if concentrations greater than 1000 suns can be incorporated. Silicon systems can be made viable (even at today's Si prices) for concentrations of 100 suns or more.

At the two American Vacuum Society Conferences in Yorktown Heights and Boston, the outlooks for low-cost thin film GaAs photovoltaics were described by the IBM Research group. Evaluations indicate that it may be possible to fabricate 10% efficient (AM1) devices using polycrystalline films of 1 μm thickness with 1-2 μm grain sizes. Even thin film single crystal devices are feasible; single crystal films of GaAs have been grown on tungsten substrates in the past.

The final session of the photovoltaics conference was devoted to CdS solar cells. A major workshop on CdS and related solar cell structures was held in Delaware during early May 1975. Advances were made in understanding the physics of these devices, in new fabrication and testing techniques, and in solving the stability problems. CdS/Cu$_2$S cells stable under open circuit conditions, and at 100°C with light incident were reported by the French group at CNES. A technique of preparing the cells in which the Cu$_2$S would lie below rather than on top of the CdS was discussed; in this way the effect of moisture and oxygen on the cells would be greatly lessened. Recent results on low-cost preparation techniques using spraying of SnO$_2$, Cu$_2$S, and CdS were described. Contributions in these areas came from the University of Stuttgart, the Universite des Science et Technique du Languedoc, the University of Delaware, CNES, and the D. H. Baldwin Company. Methods of preparing Cu$_x$S and CuInSe$_2$ on various substrates were reported by Brown University, and descriptions of nCdS/pCdTe and nCdS/nCdTe/pCu$_2$Te heterojunction cells of 4-5% efficiencies were given by Stanford University. Heterojunctions based on CdS/InP systems were reported by Bell Laboratories; these devices are about 12% efficient at AM2 and could potentially be as simple to fabricate and as low in cost as CdS/Cu$_2$S cells.

References

1. Ralph, E. L., Solar Energy *14*, 11 (1972).
2. Brown, W. C., IEEE Spectrum *10*, 38 (March 1973).
3. Wysocki, J. J., Rappaport, P., Davison, E., Hand, R., and Loferski, J. J., Appl. Phys. Lett. *9*, 44 (1966).
4. Lindmayer, J., and Allison, J. F., Conf. Rec. IEEE Photo. Spec. Conf., 9th, Silver Spring, p. 83. Also Comsat Tech. Rev. *3*, 1 (1972).
5. Mandelkorn, J., and Lamneck, J. H., Jr., Conf. Rec. IEEE Photo. Spec. Conf., 9th, Silver Spring, p. 66 (1972).
6. Gobat, A. R., Lamorte, M. F., and McIver, G. W., IRE Trans. Military Electron. *6*, 20 (1962).
7. Alferov, Zh. I., Andreev, V. M., Kagan, M. B., Protasov, I. I., and Trofim, V. G., Fiz. Tekh. Poluprov. *4*, 2378 (1970) [English Transl.: Soviet Phys.-Semicon. *4*, 2047 (1971)].
8. Woodall, J. M., and Hovel, H. J., Appl. Phys. Lett. *21*, 379 (1972).
9. Hovel, H. J., and Woodall, J. M., J. Electrochem. Soc. *120*, 1246 (1973).
10. Böer, K. W., Birchenall, C. E., Greenfield, I., Hadley, H. C., Lu, T. L., Partain, L., Phillips, J. E., Schultz, J., and Tseng, W. F., Conf. Rec. IEEE Photo. Spec. Conf., 10th, Palo Alto, p. 77 (1973).
11. Böer, K. W., Workshop Proc., Photo. Conv. Sol. Energy Terr. Appl., Cherry Hill, p. 159 (Oct. 1973). NTIS: PB-23613,23614.
12. Dash, W. C., and Newman, R., Phys. Rev. *99*, 1151 (1955).
13. Philipp, H. R., and Taft, E. A., Phys. Rev. *120*, 37 (1960).
14. Philipp, H. R., and Taft, E. A., Phys. Rev. *113*, 1002 (1959).
15. Sturge, M. D., Phys. Rev. *127*, 768 (1962).
16. Davey, J. E., and Pankey, T., J. Appl. Phys. *35*, 2203 (1964).

17. Grove, A. S., "Physics and Technology of Semiconductor Devices." Wiley, New York, 1967.
18. Ross, B., and Madigan, J. R., Phys. Rev. *108*, 1428 (1957).
19. Graff, K., Pieper, H., and Goldbach, G., "Semiconductor Silicon 1973," p. 170. Electrochem. Soc., Princeton, New Jersey, 1973.
20. Fischer, H., and Pschunder, W., Conf. Rec. IEEE Photo. Spec. Conf., 10th, Palo Alto, p. 404 (1973).
21. Iles, P. A., Conf. Rec. IEEE Photo. Spec. Conf., 8th, Seattle, p. 345 (1970).
22. Loferski, J. J., "Solar Cells, Outlook for Improved Efficiency," p. 25. National Academy of Sciences, Washington, D. C., 1972.
23. Stirn, R. J., Conf. Rec. IEEE Photo. Spec. Conf., 9th, Silver Spring, p. 72 (1972).
24. Wolf, M., Energy Conv. *11*, 63 (1971).
25. Aukerman, L. W., Millea, M. F., and McColl, M., J. Appl. Phys. *38*, 685 (1967).
26. James, L. W., Antypas, G. A., Edgecumbe, J., Moon, R. L., and Bell, R. L., J. Appl. Phys. *42*, 4976 (1971).
27. Hayashi, I., and Panish, M. B., J. Appl. Phys. *41*, 150 (1970).
28. Vilms, J., and Spicer, W. E., J. Appl. Phys. *36*, 2815 (1965).
29. James, L. W., Moll, J. L., and Spicer, W. E., "Symposium on GaAs, Dallas," p. 230. Inst. of Phys. and Phys. Soc., London, 1968.
30. Ashley, K. L., Carr, D. L., and Morano-Moran, R., Appl. Phys. Lett. *22*, 23 (1973).
31. Ettenberg, M., Kressel, H., and Gilbert, S. L., J. Appl. Phys. *44*, 827 (1973).
32. Casey, H. C., Jr., Miller, B. I., and Pinkas, E., J. Appl. Phys. *44*, 1281 (1973).
33. Ashley, K. L., and Biard, J. R., IEEE Trans. El. Dev. *ED-14*, 429 (1967).
34. Wolf, M., Proc. IEEE *51*, 674 (1963).
35. Jain, R. K., and van Overstraeten, R., J. Appl. Phys. *44*, 2437 (1973).
36. Hovel, H. J., Woodall, J. M., and Howard, W. E., "Symposium on GaAs, Boulder," p. 205. Inst. of Phys. and Phys. Soc., London, 1972.
37. Ellis, B., and Moss, T. S., Solid State Electron. *13*, 1 (1970).
38. Tsaur, S. C., Milnes, A. G., Sahai, R., and Feucht, D. L., "Symposium on GaAs, Boulder," p. 156. Inst. of Phys. and Phys. Soc., London, 1972.

39. Fossom, J. G., Sandia Laboratories, Energy Report, SLA-74-0273, June 1974.

40. Bullis, W. M., and Runyan, W. R., IEEE Trans. El. Dev. *ED-14*, 75 (1967).

41. Kaye, S., and Rolik, G. P., IEEE Trans. El. Dev. *ED-13*, 563 (1966).

42. van Overstraeten, R., and Nuyts, W., IEEE Trans. El. Dev. *ED-16*, 632 (1969).

43. Godlewski, M. P., Baraona, C. R., and Brandhorst, H. W., Jr., Conf. Rec. IEEE Photc. Spec. Conf., 10th, Palo Alto, p. 40 (1973).

44. Brandhorst, H. W., Jr., Baraona, C. R., and Swartz, C. K., Conf. Rec. IEEE Photo. Spec. Conf., 10th, Palo Alto, p. 212 (1973).

45. Hovel, H. J., and Woodall, J. M., Conf. Rec. IEEE Photo. Spec. Conf., 10th, Palo Alto, p. 25 (1973).

46. Faraday, B. J., Statler, R. L., and Tauke, R. V., Proc. IEEE *56*, 31 (1968).

47. Mytton, R. J., Brit. J. Appl. Phys. Ser. 2, *1*, 721 (1968).

48. Wolf, M., Conf. Rec. IEEE Photo. Spec. Conf., 9th, Silver Spring, p. 53 (1972).

49. "Solar Electromagnetic Radiation." NASA Bull. SP-8005, revised May 1971.

50. Thekaekara, M. P., Solar Energy *14*, 109 (1973).

51. Moon, P., J. Franklin Inst. *230*, 583 (1940).

52. Yasui, R. K., and Schmidt, L. W., Conf. Rec. IEEE Photo. Spec. Conf., 8th, Seattle, p. 110 (1970).

53. Gaddy, E. M., Conf. Rec. IEEE Photo. Spec. Conf., 10th, Palo Alto, p. 153 (1973).

54. Bogus, K., and Mattes, S., Conf. Rec. IEEE Photo. Spec. Conf., 9th, Silver Spring, p. 106 (1972).

55. Fahrenbruch, A. L., and Bube, R. H., Conf. Rec. IEEE Photo. Spec. Conf., 10th, Palo Alto, p. 85 (1973).

56. Böer, K. W., and Phillips, J., Conf. Rec. IEEE Photo. Spec. Conf., 9th, Silver Spring, p. 125 (1972).

57. Sah, C. T., Noyce, R. N., and Shockley, W., Proc. IRE *45*, 1228 (1957).

58. Choo, S. C., Solid State Electron. *11*, 1069 (1968).

59. Hovel, H. J., Conf. Rec. IEEE Photo. Spec. Conf., 10th, Palo Alto, p. 34 (1973).

60. Milnes, A. G., and Feucht, D. L., "Heterojunctions and Metal-Semiconductor Junctions." Academic Press, New York, 1972.

61. Hovel, H. J., A Review of the Principles of Semiconductor Heterojunctions, IBM Research Reports RC 2786 (1970).

62. Stirn, R. J., private communication.

63. Wolf, M., and Rauschenback, Adv. Energy Conv. *3*, 455 (1963).
64. Loferski, J. J., Acta Electron. *5*, 350 (1961).
65. Loferski, J. J., Proc. IEEE *51*, 667 (1963).
66. Lindmayer, J., Comsat Tech. Rev. *2*, 105 (1972).
67. Handy, R. J., Solid State Electron. *10*, 765 (1967).
68. Wolf, M., Proc. IRE *48*, 1246 (1960).
69. Loferski, J. J., Ranganathan, N., Crisman, E. E., and Chen, L. Y., Conf. Rec. IEEE Photo. Spec. Conf., 9th, Silver Spring, p. 19 (1972).
70. Sah, C. T., IRE Trans. El. Dev. *ED-9*, 94 (1962).
71. Shockley, W., and Queisser, H. J., J. Appl. Phys. *32*, 510 (1961).
72. Shockley, W., and Henley, R., Bull. Am. Phys. Soc. *6*, 106 (1961).
73. Nakamura, M., Kato, T., and Oi, N., Jpn. J. Appl. Phys. *7*, 512 (1968).
74. Gill, W. D., and Bube, R. H., J. Appl. Phys. *41*, 3731 (1970).
75. Lindquist, P. F., and Bube, R. H., J. Appl. Phys. *43*, 2839 (1972).
76. Lindquist, P. F., and Bube, R. H., J. Electrochem. Soc. *119*, 936 (1972).
77. Martinuzzi, S., Cabane-Brouty, F., and Bretzner, J. F., Conf. Rec. IEEE Photo. Spec. Conf., 9th, Silver Spring, p. 111 (1972). Also Martinuzzi, S., and Mallem, O., Phys. Status Solidi *16*, 339 (1973).
78. Fahrenbruch, A. L., and Bube, R. H., J. Appl. Phys. *45*, 1264 (1974).
79. Kendall, D., Conf. Phys. Appl. Li-Diffused Si, NASA-Goddard, Dec. 1969.
80. Sze, S. M., "Physics of Semicon. Devices," Chapter 2. Wiley, New York, 1969.
81. Sell, D. D., and Casey, H. C., Jr., J. Appl. Phys. *45*, 800 (1974).
82. Berman, P. A., Conf. Rec. IEEE Photo. Spec. Conf., 9th, Silver Spring, p. 281 (1972).
83. Iles, P. A., Conf. Rec. IEEE Photo. Spec. Conf., 9th, Silver Spring, p. 296 (1972).
84. Huber, D., and Bogus, K., Conf. Rec. IEEE Photo. Spec. Conf., 10th, Palo Alto, p. 100 (1973).
85. Palz, W., Besson, J., Nguyen Duy, T., and Vedel, J., Conf. Rec. IEEE Photo. Spec. Conf., 10th, Palo Alto, p. 69 (1973).
86. Wolf, M., and Ralph, E. L., IEEE Trans. El. Dev. *ED-12*, 470 (1965).

87. Iles, P. A., and Zemmrich, D. K., Conf. Rec. IEEE Photo.
     Spec. Conf., 10th, Palo Alto, p. 200 (1973).
88. Mlavsky, A. I., Proc. Symp. Mat. Sci. Aspects Thin Films
     Solar Energy Conv., Tucson (May 1974). NTIS: NSF-RA-
     N-74-062.
89. Chu, T. L., Proc. Symp. Mat. Sci. Aspects Thin Films
     Solar Energy Conv., Tucson (May 1974). Also, Workshop
     Proc., Photo. Conv. Solar Energy Terr. Appl., Cherry
     Hill, p. 56 (Oct. 1973).
90. Mandelkorn, J., Lamneck, J. H., and Scudder, L. R.,
     Conf. Rec. IEEE Photo. Spec. Conf., 10th, Palo Alto,
     p. 207 (1973).
91. Redfield, D., Appl. Phys. Lett. *25*, 647 (1974).
92. Shockley, W., "Electrons and Holes in Semiconductors,"
     p. 318. Van Nostrand, Princeton, New Jersey, 1950.
93. Rai-Choudhuri, P., and Hower, P. L., J. Electrochem.
     Soc. *120*, 1761 (1973).
94. Ettenberg, M., J. Appl. Phys. *45*, 901 (1974).
95. Soclof, S., and Iles, P. A., Extended Abstracts, Elec-
     trochem. Soc. Fall Meeting, New York, p. 618 (1974).
96. Heaps, J. D., Tufte, O. N., and Nussbaum, A., IEEE
     Trans. El. Dev. *ED-8*, 560 (1961).
97. Fang, P. H., Workshop Proc., Photo. Conv. Solar Energy
     Terr. Appl., Cherry Hill, p. 51 (Oct. 1973). Also
     Fang, P. H., Ephrath, L., and Nowak, W. B. (1974), Appl.
     Phys. Lett. *25*, 583 (1974).
98. Berry, W. B., Workshop Proc., Photo. Conv. Solar Energy
     Terr. Appl., Cherry Hill, p. 67 (Oct. 1973).
99. Iles, P. A., Workshop Proc., Photo. Conv. Solar Energy
     Terr. Appl., Cherry Hill, p. 71 (Oct. 1973).
100. Vohl, P., Perkins, D. M., Ellis, S. G., Addiss, R. R.,
     Hui, W., and Noel, G., IEEE Trans. El. Dev. *ED-14*, 26
     (1967).
101. Dutton, D., Phys. Rev. *112*, 785 (1958).
102. Mead, C. A., Solid State Electron. *9*, 1023 (1966).
103. Sze, S. M., "Physics of Semiconductor Devices," Chap-
     ter 8. Wiley, New York, 1969.
104. Card, H. C., and Rhoderick, E. H., J. Phys. D: Appl.
     Phys. *4*, 1589 (1971).
105. Li, S. S., Lindholm, F. A., and Wang, C. T., J. Appl.
     Phys. *43*, 4123 (1972).
106. Schneider, M. V., Bell System Tech. J. *45*, 1611 (1966).
107. Stirn, R. J., and Yeh, Y. M., Conf. Rec. IEEE Photo.
     Spec. Conf., 10th, Palo Alto, p. 15 (1973).
108. Baertsch, R. D., and Richardson, J. R., J. Appl. Phys.
     40, 229 (1969).

109. Rhoderick, E. H., J. Phys. D: Appl. Phys. *3*, 1153 (1970).

110. Andrews, J. M., and Lepselter, M. P., Solid State Electron. *13*, 1011 (1970).

111. Crowell, C. R., and Sze, S. M., Solid State Electron. *9*, 1035 (1966).

112. Smith, B. L., and Rhoderick, E. H., Solid State Electron. *14*, 71 (1971).

113. Chang, C. Y., and Sze, S. M., Solid State Electron. *13*, 727 (1970).

114. Anderson, W. A., Milano, R. A., Delahoy, A. E., and Vernon, S., Extended Abstracts, Electrochem. Soc. Fall Meeting, New York, p. 621 (1974).

115. Pulfrey, D. L., and McOuat, R. F., Appl. Phys. Lett. *24*, 167 (1974).

116. Anderson, W. A., Delahoy, A. E., and Milano, R. A., J. Appl. Phys. *45*, 3913 (1974).

117. Riben, A. R., and Feucht, D. L., Internat. J. Electron. *20*, 583 (1966).

118. Jadus, D. K., and Feucht, D. L., IEEE Trans. El. Dev. *ED-16*, 102 (1969).

119. Hovel, H. J., and Milnes, A. G., IEEE Trans. El. Dev. *ED-16*, 766 (1969).

120. Dumke, W. P., Woodall, J. M., and Rideout, V. L., Solid State Electron. *15*, 1339 (1972).

121. Alferov, Zh. I., Andreev, V. M., Korol'Kov, V. I., Portnoi, E. L., and Tret'yakov, D. N., Fiz. Tekh. Poluprov. *4*, 167 (1970) [English transl.: Sov. Phys. Semicon. *4*, 132 (1970)].

122. Sreedhar, A. K., Sharma, B. L., and Purohit, R. K., IEEE Trans. El. Dev. *ED-16*, 309 (1969).

123. Sahai, R., and Milnes, A. G., Solid State Electron. *13*, 1289 (1970).

124. Alferov, Zh. I., Zimorgorova, N. S., Trukan, M. K., and Tuchkevich, V. M., Fiz. Tver. Tela *7*, 1235 (1965) [English transl.: Sov. Phys. Solid State *7*, 990 (1965)].

125. Purohit, R. K., Phys. Status Solidi *24*, K57 (1967).

126. Okimura, H., Kawakami, M., and Sakai, Y., Jpn. J. Appl. Phys. *6*, 908 (1967).

126a. Justi, E. W., Schneider, G., and Seredynski, J., Energy Conv. *13*, 53 (1973).

127. Cusano, D. A., Solid State Electron. *6*, 217 (1963).

127a. Fahrenbruch, A. L., Vasilchenko, V., Buch, F., Mitchell, K., and Bube, R. H., Appl. Phys. Lett. *25*, 605 (1974).

127b. Wagner, S., Shay, J. L., Migliorato, P., and Kasper, H. M., Appl. Phys. Lett. *25*, 434 (1974); Appl. Phys. Lett. *27*, 89 (1975).

127c. Wagner, S., Shay, J. L., Bachmann, K. J., and Buehler,
      E., Appl. Phys. Lett. *26*, 229 (1975).
128. Rahilly, W. P., Conf. Rec. IEEE Photo. Spec. Conf., 9th,
      Silver Spring, p. 44 (1972).
129. Stella, P., and Gover, A., Conf. Rec. IEEE Photo. Spec.
      Conf., 9th, Silver Spring, p. 85 (1972).
130. Gover, A., and Stella, P., IEEE Trans. El. Dev. *ED-21*,
      351 (1974).
131. Chadda, T. B. S., and Wolf, M., Conf. Rec. IEEE Photo.
      Spec. Conf., 9th, Silver Spring, p. 87 (1972).
132. Chadda, T. B. S., and Wolf, M., Conf. Rec. IEEE Photo.
      Spec. Conf., 10th, Palo Alto, p. 52 (1973).
133. Sater, B. L., Brandhorst, H. W., Jr., Riley, T. J., and
      Hart, R. E., Jr., Conf. Rec. IEEE Photo. Spec. Conf.,
      10th, Palo Alto, p. 188 (1973).
134. Smeltzer, R. K., Kendall, D. L., and Varnell, G. L.,
      Conf. Rec. IEEE Photo. Spec. Conf., 10th, Palo Alto,
      p. 194 (1973).
135. Loferski, J. J., Crisman, E. E., Armitage, W., and Chen,
      L. Y., Conf. Rec. IEEE Photo. Spec. Conf., 10th, Palo
      Alto, p. 58 (1973).
136. Wang, C. T., and Li, S. S., IEEE Trans. El. Dev. *ED-20*,
      522 (1973).
137. Hess, W. N., "The Radiation Belt and Magnetosphere."
      Xerox College Publishing, New York, 1968.
138. O'Brien, B. J., "Radiation Belts," p. 84. Scientific
      American, May 1963.
139. Loferski, J. J., and Rappaport, P., J. Appl. Phys. *30*,
      1181 (1959).
140. Rappaport, P., and Wysocki, J. J., Acta Electron. *5*,
      364 (1961).
141. Baicker, J. A., and Faughnan, B. W., J. Appl. Phys. *33*,
      3271 (1962).
142. Smits, F. M., IEEE Trans. Nucl. Sci. *NS-10*, 88 (1963).
143. Burrill, J. T., King, W. J., Harrison, S., and McNally,
      P., IEEE Trans. El. Dev. *ED-14*, 10 (1967).
144. Meulenberg, A., Jr., and Treble, F. C., Conf. Rec. IEEE
      Photo. Spec. Conf., 10th, Palo Alto, p. 359 (1973).
145. Rosenzweig, W., Bell System Tech. J. *41*, 1573 (1962).
146. Wilsey, N. D., Conf. Rec. IEEE Photo. Spec. Conf., 9th,
      Silver Spring, p. 338 (1972).
147. Rostron, R. W., Energy Conv. *12*, 125 (1972).
148. Wysocki, J. J., Rappaport, P., Davison, E., and Loferski,
      J. J., IEEE Trans. El. Dev. *ED-13*, 420 (1966).
149. Mandelkorn, J., Schwartz, L., Broder, J., Kautz, H., and
      Ulman, R., J. Appl. Phys. *35*, 2258 (1964).
150. Srour, J. R., Othmer, S., and Curtis, O. L., Jr., Conf.

Rec. IEEE Photo. Spec. Conf., 9th, Silver Spring, p. 336 (1972).

151. Crabb, R. L., Conf. Rec. IEEE Photo. Spec. Conf., 10th, Palo Alto, p. 396 (1973).

152. Lindmayer, J., and Arndt, R. A., Conf. Rec. IEEE Photo. Spec. Conf., 10th, Palo Alto, p. 358 (1973).

153. Wallis, A. E., and Green, J. M., Conf. Rec. IEEE Photo. Spec. Conf., 10th, Palo Alto, p. 373 (1973).

154. Curtin, D. J., and Cool, R. W., Conf. Rec. IEEE Photo. Spec. Conf., 10th, Palo Alto, p. 139 (1973).

155. Wysocki, J. J., IEEE Trans. Nucl. Sci. *NS-14,* 103 (1967).

156. Young, R. C., Westhead, J. W., and Corelli, J. C., J. Appl. Phys. *40,* 271 (1969).

157. Reynard, D. L., and Peterson, D. G., Conf. Rec. IEEE Photo. Spec. Conf., 9th, Silver Spring, p. 303 (1972).

158. Faith, T. J., Conf. Rec. IEEE Photo. Spec. Conf., 9th, Silver Spring, p. 292 (1972).

159. Anspaugh, B. E., and Carter, J. R., Conf. Rec. IEEE Photo. Spec. Conf., 10th, Palo Alto, p. 366 (1973).

160. Godlewski, M. P., Baraona, C. R., and Brandhorst, H. W., Jr., Conf. Rec. IEEE Photo. Spec. Conf., 10th, Palo Alto, p. 378 (1973).

161. Fang, P. H., and Liu, Y. M., Phys. Lett. *20,* 344 (1966).

162. Goldhammer, L. J., and Anspaugh, B. E., Conf. Rec. IEEE Photo. Spec. Conf., 8th, Seattle, p. 201 (1970).

163. Faile, S. P., Harding, W. R., and Wallis, A. E., Conf. Rec. IEEE Photo. Spec. Conf., 8th, Seattle, p. 88 (1970).

164. Forestieri, A. F., and Broder, J. D., Conf. Rec. IEEE Photo. Spec. Conf., 8th, Seattle, p. 179 (1970).

165. Broder, J. D., and Mazaris, G. A., Conf. Rec. IEEE Photo. Spec. Conf., 10th, Palo Alto, p. 272 (1973).

166. Crabb, R. L., Conf. Rec. IEEE Photo. Spec. Conf., 9th, Silver Spring, p. 185 (1972).

167. Kirkpatrick, A. R., Tripoli, G. A., and Bartels, F. T. C., Conf. Rec. IEEE Photo. Spec. Conf., 8th, Seattle, p. 176 (1970).

168. Brackley, G., Lawson, K., and Satchell, D. W., Conf. Rec. IEEE Photo. Spec. Conf., 9th, Silver Spring, p. 174 (1972).

169. Stella, P. M., and Somberg, H., Conf. Rec. IEEE Photo. Spec. Conf., 9th, Silver Spring, p. 179 (1972).

170. Rauch, H. W., Sr., Ulrich, D. R., and Green, J. M., Conf. Rec. IEEE Photo. Spec. Conf., 10th, Palo Alto, p. 182 (1973).

171. Runyan, W. R., and Alexander, E. A., IEEE Trans. El. Dev. *ED-14,* 3 (1967).

172. Wysocki, J. J., J. Appl. Phys. *34*, 2915 (1963).
173. van Aerschodt, A. E., Capart, J. J., David, K. H., Fab-
      bricotti, M., Heffels, K. H., Loferski, J. J., and
      Reinhartz, K. K., IEEE Trans. El. Dev. *ED-18*, 471 (1971).
174. Mandelkorn, J., Baraona, C. R., and Lamneck, J. H., Jr.,
      Conf. Rec. IEEE Photo. Spec. Conf., 9th, Silver Spring,
      p. 15 (1972).
175. Panish, M. B., and Casey, H. C., Jr., J. Appl. Phys. *40*,
      163 (1969).
176. Luft, W., IEEE Trans. Aerospace Electron. Systems *AES-7*,
      332 (1971).
177. Hovel, H. J., and Woodall, J. M., Quarterly Progress
      Report, NASA Contract 12812, 1 October, 1974.
177a. Vernon, S. M., and Anderson, W. A., Appl. Phys. Lett.
      26, 707 (1975).
178. Luft, W., IEEE Trans. Aerospace Electron. Systems *AES-6*,
      797 (1970).
179. Davis, R., and Knight, J. R. (unpublished).
179a. Vasil'ev, A. M., Evdokimov, V. M., Landsman, A. P., and
      Milovanov, A. F., Geliotekh. *11*, 18 (1975) [English
      transl.: Appl. Sol. Energy *11*, 72 (1975)].
180. Magee, V., Webb, H. G., Haigh, A. D., and Freestone, R.,
      Conf. Rec. IEEE Photo. Spec. Conf., 9th, Silver Spring,
      p. 6 (1972).
181. Payne, P. A., and Ralph, E. L., Conf. Rec. IEEE Photo.
      Spec. Conf., 8th, Seattle, p. 135 (1970).
182. Brandhorst, H. W., Jr., and Hart, R. E., Jr., Conf. Rec.
      IEEE Photo. Spec. Conf., 8th, Seattle, p. 142 (1970).
183. Ho, J. C., Bartels, F. T. C., and Kirkpatrick, A. R.,
      Conf. Rec. IEEE Photo. Spec. Conf., 8th, Seattle, p. 150
      (1970).
184. Rhodes, R. G., "Imperfections and Active Centers in
      Semiconductors." Macmillan, New York, 1964.
185. Bates, H. E., Cocks, F. H., and Mlavsky, A. I., Conf.
      Rec. IEEE Photo. Spec. Conf., 9th, Silver Spring, p. 386
      (1972).
185a. Ciszek, T. F., Mat. Res. Bull. 7, 731 (1972).
186. Surek, T., and Chalmers, B., Workshop Proc., Photo. Conv.
      Solar Energy Terr. Appl., Cherry Hill, p. 13 (1973).
187. Mlavsky, A. I., Workshop Proc., Photo. Conv. Solar
      Energy Terr. Appl., Cherry Hill, p. 22 (1973).
188. Bates, H. E., Jewett, D. N., and White, V. E., Conf.
      Rec. IEEE Photo. Spec. Conf., 10th, Palo Alto, p. 197
      (1973).
189. Dermatis, S. N., Faust, J. W., Jr., John, H. F., J.
      Electrochem. Soc. *112*, 792 (1965).

189a. Tucker, T. N., and Schwuttke, G. H., Appl. Phys. Lett. *9*, 219 (1966).
189b. Tucker, T. N., and Hood, J. S., Electrochem. Tech. *6*, 49 (1968).
189c. Barrett, D. L., Myers, E. H., Hamilton, D. R., and Bennett, A. I., J. Electrochem. Soc. *118,* 952 (1971).
190. Currin, C. G., Ling, K. S., Ralph, E. L., Smith, W. A., and Stirn, R. J., Conf. Rec. IEEE Photo. Spec. Conf., 9th, Silver Spring, p. 363 (1972).
191. Mlavsky, A. I., Symp. Mat. Sci. Aspects Thin Films Solar Energy Conv., Tucson (May 1974).
192. Kressel, H., Robinson, P., McFarlane, S. H., D'Aiello, R. V., and Dalal, V. L., Appl. Phys. Lett. *25,* 197 (1974).
193. Nowak, M. B., and Fang, P. H., Workshop Proc., Photo. Conv. Solar Energy Terr. Appl., Cherry Hill, p. 54 (Oct. 1973).
194. Chu, T. L., Proc. Symp. Mat. Sci. Aspects Thin Films Solar Energy Conv., Tucson (May 1974).
195. Hall, L. H., and Koliwad, K. M., J. Electrochem. Soc. *120,* 1438 (1973).
196. Yasuda, Y., and Moriya, T., "Semiconductor Silicon 1973," p. 271. Electrochemical Society, Princeton, New Jersey, 1973.
197. Chiang, Y. S., "Semiconductor Silicon 1973," p. 285. Electrochemical Society, Princeton, New Jersey, 1973.
198. Ban, Y., Tsuchikawa, H., and Maeda, K., "Semiconductor Silicon 1973," p. 292. Electrochemical Society, Princeton, New Jersey, 1973.
199. Celotti, G., Nobili, D., and Ostoja, P., J. Mat. Sci. *9,* 821 (1974).
200. Mandelkorn, J., McAffee, C., Kesperis, J., Schwartz, L., and Pharo, W., J. Electrochem. Soc. *109,* 313 (1962).
201. Faith, T. J., Corra, J. P., and Holmes-Siedle, A. G., Conf. Rec. IEEE Photo. Spec. Conf., 8th, Seattle, p. 247 (1970).
202. Carter, J. R., and Downing, R. G., Conf. Rec. IEEE Photo. Spec. Conf., 8th, Seattle, p. 240 (1970).
203. Tannenbaum, E., Solid State Electron. *2,* 123 (1961).
204. McDonald, R. A., Ehlenberger, G. G., and Huffman, T. R., Solid State Electron. *9,* 807 (1966).
205. Tsai, J. C. C., Proc. IEEE *57,* 1499 (1969).
206. Lamneck, J. H., Jr., Schwartz, L., and Spakowski, A. E., Conf. Rec. IEEE Photo. Spec. Conf., 9th, Silver Spring, p. 193 (1972).
207. Kamins, T. I., J. Electrochem. Soc. *121,* 286 (1974).

208. Lever, R. F., and Demsky, H. M., IBM J. Res. Dev. *18*, 40 (1974).

209. Steinemann, A., and Zimmerli, U., "Crystal Growth" (H. S. Peifer, ed.), p. 81. Pergamon, New York, 1967.

210. Plaskett, T. S., Woodall, J. M., and Segmüller, A., J. Electrochem. Soc. *118*, 115 (1971).

211. Goundry, P. C., "Symposium on GaAs (Proceedings)," Reading, p. 31. Inst. of Phys. and Phys. Soc., London, 1966.

212. Berkowitz, J. B., Workshop Proc., Photo. Conv. Solar Energy Terr. Appl., Cherry Hill, p. 232 (Oct. 1973).

213. Jain, V. K., and Sharma, S. K., Solid State Electron. *13*, 1145 (1970).

214. Bradshaw, A., and Knappett, J. E., Solid State Tech. *13*, 45 (1970).

215. Manasevit, H. M., and Simpson, W. I., J. Electrochem. Soc. *116*, 1725 (1969).

216. Rai-Choudhury, P., J. Electrochem. Soc. *116*, 1745 (1969).

217. Blakeslee, A. E., and Bischoff, B. K., Electrochemical Society Fall Meeting, Cleveland, Abstract #181 (1971).

218. Nelson, H., RCA Rev. *24*, 603 (1963).

219. Various papers in "Symposium on GaAs (Proceedings)," Dallas, Oct. 1968, and "Symposium on GaAs (Proceedings)," Boulder, Sept. 1972. Inst. of Phys. and Phys. Soc., London, 1968, 1972.

220. Willardson, R. K., and Allred, W. P., "Symposium on GaAs (Proceedings)," Reading, p. 35. Inst. of Phys. and Phys. Soc., London, 1966.

221. Kagan, M. B., Landsman, A. P., and Kholev, B. A. Fiz. Tekh. Poluprov. *1*, 918 (1967) [English transl.: Sov. Phys. Semicon. *1*, 761 (1967)].

222. Jenny, D. A., Loferski, J. J., and Rappaport, P., Phys. Rev. *101*, 1208 (1956).

223. Casey, H. C., Jr., "Atomic Diffusion in Semiconductors" D. Shaw, ed.), Chapter 6. Plenum, New York, 1973. Also Casey, H. C., Jr., and Panish, M. B., Trans. Met. Soc. AIME *242*, 406 (1968).

224. Marinace, J. C., IBM J. Res. Dev. *15*, 258 (1971).

225. Andreev, V. M., Golovner, T. M., Kagan, M. B., Koroleva, N. S., Lyubashevskaya, T. L., Nuller, T. A., and Tret'yakov, D. N. Fiz. Tekh. Poluprov. *7*, 2289 (1973) [English transl.: Sov. Phys. Semicon. *7*, 1525 (1974)].

226. Spakowski, A. E., IEEE Trans. El. Dev. *ED-14*, 18 (1967).

227. Mytton, R. J., Clark, L., Gale, R. W., and Moore, K., Conf. Rec. IEEE Photo. Spec. Conf., 9th, Silver Spring, p. 133 (1972).

228. Palz, W., Besson, J., Nguyen Duy, T., and Vedel, J.,
      Conf. Rec. IEEE Photo. Spec. Conf., 9th, Silver Spring,
      p. 91 (1972).
229. Bernatowicz, D. T., and Brandhorst, H. W., Jr., Conf.
      Rec. IEEE Photo. Spec. Conf., 8th, Seattle, p. 24 (1970).
230. Palz, W., Besson, J., Fremy, J., Nguyen Duy, T., and
      Vedel, J., Conf. Rec. IEEE Photo. Spec. Conf., 8th,
      Seattle, p. 16 (1970).
231. Mathieu, H. J., Reinhartz, K. K., and Rickert, H., Conf.
      Rec. IEEE Photo. Spec. Conf., 10th, Palo Alto, p. 93
      (1973).
232. Palz, W., Besson, J., Nguyen Duy, T., and Vedel, J.,
      Conf. Rec. IEEE Photo. Spec. Conf., 10th, Palo Alto,
      p. 69 (1973).
232a. Mytton, R. J., Solar Energy *16*, 33 (1974).
233. Brody, T. P., and Shirland, F. A., Workshop Proc.,
      Photo. Conv. Solar Energy Terr. Appl., Cherry Hill,
      p. 232 (Oct. 1973).
234. Jordan, J. F., Workshop Proc., Photo. Conv. Solar Energy
      Terr. Appl., Cherry Hill, p. 182 (1973).
235. Balch, J. W., and Anderson, W. W., Phys. Status Solidi.
      9, 567 (1972).
236. Baczewski, A., J. Electrochem. Soc. *112*, 577 (1965).
237. Parker, S. G., Pinnell, J. E., and Swink, L. N., J.
      Phys. Chem. Solids *32*, 139 (1971).
238. Lilley, P., Jones, P. L., and Litting, C. N. W., J.
      Mat. Sci. *5*, 891 (1970).
239. Igarashi, O., J. Appl. Phys. *41*, 3190 (1970).
240. Noack, J., and Möhling, W., Phys. Status Solidi (a) *3*,
      K229 (1970).
241. Thomas, R. W., J. Electrochem. Soc. *116*, 1449 (1969).
242. Rosztoczy, F. E., and Stein, W. W., J. Electrochem.
      Soc. *119*, 1119 (1972).
243. Fraser, D. B., and Cook, H. D., J. Electrochem. Soc.
      119, 1368 (1972).
244. Nishino, T., and Hamakawa, Y., Japan. J. Appl. Phys. *9*,
      1085 (1972).
245. Fillard, J. P., and Manifacier, J. C., Japan. J. Appl.
      Phys. *9*, 1012 (1970).
246. Mehta, R. R., and Vogel, S. F., J. Electrochem. Soc.
      119, 752 (1972).
247. Kajiyama, K., and Furukawa, Y., Japan. J. Appl. Phys.
      6, 905 (1967).
248. Nishino, T., and Hamakawa, Y., Proc. Int. Conf. Phys.
      Chem. Semicon. Hetjns, Budapest, Vol. II, p. 409,
      sponsored by Intern. Union of Pure and Appl. Phys. and
      the European Phys. Soc. (1972).

249. Cunnell, F. A., and Gooch, C. H., J. Phys. Chem. Solids *15*, 127 (1960).
250. Gusev, V. M., Zaddé, V. V., Landsman, A. P., and Titov, V. V., Fiz. Tver. Tela. *8*, 1708 (1966) [English transl.: Sov. Phys. Sol. State *8*, 1363 (1966)].
251. "Development of GaAs Solar Cells." Final Report, Contract 953270, Ion Physics Corp., Feb. 1973. (NASA-CR-135510).
252. Vaidyanathan, K. V., and Walker, G. H., Conf. Rec. IEEE Photo. Spec. Conf., 10th, Palo Alto, p. 31 (1973).
253. "Encyclopedia of Chemical Technology" (Kirk-Othmer, eds.) Vol. 18. Wiley, New York, 1964.
254. Seraphin, B. O., and Bennett, H. E., "Semiconductors and Semimetals" (Willardson and Beer, eds.), Vol. 3. Academic Press, New York, 1967.
255. "Thermophysical Properties of Matter," Vol. 8, Non-metallic Solids. IFI Plenum, New York, 1972.
256. Wang, E. Y., Yu, F. T. S., Simms, V. L., Brandhorst, H. W., Jr., and Broder, J. D., Conf. Rec. IEEE Photo. Spec. Conf., 10th, Palo Alto, p. 168 (1973).
257. Anders, H., "Thin Films in Optics," Chapter 1. Focal Press, London, 1967.
258. Crabb, R. L., and Atzei, A., Conf. Rec. IEEE Photo. Spec. Conf., 8th, Seattle, p. 78 (1970).
259. Schwartz, J. P., Conf. Rec. IEEE Photo. Spec. Conf., 8th, Seattle, p. 173 (1970).
260. Musset, A., and Thelen, A., "Progress in Optics" (E. Wolf, ed.), Chapter 4. North-Holland, Amsterdam, 1970.
261. Knausenberger, W. H., and Tauber, R. N., J. Electrochem. Soc. *120*, 927 (1973).
262. Revesz, A. G., Conf. Rec. IEEE Photo. Spec. Conf., 10th, Palo Alto, p. 180 (1973).
263. Roger, J., and Colardelle, P., Conf. Rec. IEEE Photo. Spec. Conf., 8th, Seattle, p. 84 (1970).
264. Science News, p. 263, Oct. 26, 1974.
265. Fischer, H., and Gereth, R., IEEE Trans. Elect. Dev. *ED-18*, 459 (1971).
266. Becker, W. H., and Pollack, S. R., Conf. Rec. IEEE Photo. Spec. Conf., 8th, Seattle, p. 40 (1970).
267. Liu, K., and Yasui, R. K., Conf. Rec. IEEE Photo. Spec. Conf., 8th, Seattle, p. 62 (1970).
268. Braslau, N., Gunn, J. B., and Staples, J. L., Solid State Electron. *10*, 381 (1967).
269. Killesreiter, H., and Baessler, H., Chem. Phys. Lett. *11*, 411 (1971).
270. Ghosh, A. K., and Feng, T., J. Appl. Phys. *44*, 2781 (1973).

271. Usov, N. N., and Benderskii, V. A., Fiz. Tekh. Poluprov.
     2, 699 (1968) [English transl.: Sov. Phys. Semicon. *2*,
     580 (1968)].

272. Fedorov, M. I., and Benderskii, V. A., Fiz. Tekh. Poluprov.
     4, 1403 (1970) [English transl.: Sov. Phys. Semicon. *4*,
     1198 (1971)].

273. Fedorov, M. I., and Benderskii, V. A., Fiz. Tekh. Poluprov.
     4, 2007 (1970) [English transl.: Sov. Phys. Semicon. *4*,
     1720 (1971)].

274. Ghosh, A. K., Morel, D. L., Feng, T., Shaw, R. F., and
     Rowe, C. A., Jr., J. Appl. Phys. *45*, 230 (1974).

275. Tang, C. W., and Albrecht, A. C., J. Chem. Phys., *62*,
     2139 (1975); also Tang, C. W., and Albrecht, A. C.,
     Nature, to be published.

276. Tang, C. W., and Albrecht, A. C., to be submitted.

277. Lyons, L. E., and Newman, O. M. G., Aust. J. Chem. *24*,
     13 (1971).

278. Hovel, H. J., and Milnes, A. G., Int. J. Electron. *25*,
     201 (1968).

279. Reucroft, P. J., Takahashi, K., and Ullal, H., Appl.
     Phys. Lett. *25*, 664 (1974).

280. Wolf, M., Conf. Rec. IEEE Photo. Spec. Conf., 9th,
     Silver Spring, p. 342 (1972).

281. Böer, K. W., Conf. Rec. IEEE Photo. Spec. Conf., 9th,
     Silver Spring, p. 351 (1972).

282. Spakowski, A. E., and Shure, L., Conf. Rec. IEEE Photo.
     Spec. Conf., 9th, Silver Spring, p. 359 (1972).

283. Wolf, M., Conf. Rec. IEEE Photo. Spec. Conf., 10th,
     Palo Alto, p. 5 (1973).

284. Wolf, M., Energy Conv. *14*, 9 (1974).

285. Iles, P. A., Proc. Symp. Mat. Sci. Aspects Thin Films
     Solar Energy Conv., Tucson, p. 37 (May 1974).

286. "McGraw-Hill Encyclopedia of Science and Technology,"
     Vol. 4, p. 550. McGraw-Hill, New York, 1966.

287. "United States Mineral Resources" (D. A. Brobst and
     W. P. Pratt, eds.). U.S. Dept. of the Interior, Geo-
     logical Survey Professional Paper 820, U.S.G.P.O.,
     Washington, D. C., 1973.

288. "Assessment of the Tech. Required to Dev. Photovolt.
     Power Systs. for Large Scale National Energy Appl.,"
     JPL Special Publication 43-11, Oct. 15, 1974 (NSF-RA-
     N-74-072).

289. James, L. W., and Moon, R. L., Varian Corporate Research
     Memorandum CRM-286, Nov. 5, 1974. See also Chapter VIII.

290. Auvergne, D., Camassel, J., and Mathieu, H., Phys. Rev. B *11*, 2251 (1975).

290a. Serreze, H. B., Swartz, J. C., and Entine, G., Mater. Res. Bull. *9*, 1421 (1974).

291. Ouwens, C. D., and Heijligers, H., Appl. Phys. Lett. *26*, 569 (1975).

292. Anderson, W. A., and Milano, R. A., Proc. IEEE *63*, 206 (1975); also Vernon, S. M., and Anderson, W. A., Appl. Phys. Lett. *26*, 707 (1975).

293. Charlson, E., and Lien, C., J. Appl. Phys., to be published.

294. James, L. W., and Moon, R. L., Appl. Phys. Lett. *26*, 467 (1975).

295. Davis, R., and Knight, J. R., Solar Energy *17*, 145 (1975).

Index